Solid Mechanics and Its Applications

Volume 204

Series Editor

G. M. L. Gladwell
Department of Civil Engineering, University of Waterloo, Waterloo, ON, Canada

For further volumes:
http://www.springer.com/series/6557

Aims and Scope of the Series

The fundamental questions arising in mechanics are: *Why*?, *How*?, *and How much*? The aim of this series is to provide lucid accounts written by authoritative researchers giving vision and insight in answering these questions on the subject of mechanics as it relates to solids.

The scope of the series covers the entire spectrum of solid mechanics. Thus it includes the foundation of mechanics; variational formulations; computational mechanics; statics, kinematics and dynamics of rigid and elastic bodies: vibrations of solids and structures; dynamical systems and chaos; the theories of elasticity, plasticity and viscoelasticity; composite materials; rods, beams, shells and membranes; structural control and stability; soils, rocks and geomechanics; fracture; tribology; experimental mechanics; biomechanics and machine design.

The median level of presentation is the first year graduate student. Some texts are monographs defining the current state of the field; others are accessible to final year undergraduates; but essentially the emphasis is on readability and clarity.

P. Podio-Guidugli · A. Favata

Elasticity for Geotechnicians

A Modern Exposition of Kelvin, Boussinesq, Flamant, Cerruti, Melan, and Mindlin Problems

Springer

P. Podio-Guidugli
Department of Civil Engineering
and Information Engineering
University of Rome Tor Vergata
Rome
Italy

A. Favata
Institute of Continuum Mechanics
and Material Mechanics
Hamburg University of Technology
Hamburg
Germany

ISSN 0925-0042
ISBN 978-3-319-34530-7 ISBN 978-3-319-01258-2 (eBook)
DOI 10.1007/978-3-319-01258-2
Springer Cham Heidelberg New York Dordrecht London

Printed on acid-free paper

Springer is part of Springer Science+Business Media (www.springer.com)

About this Book

This book deals in a modern manner with a family of problems from an old and mature subject, *classical elasticity*.

Classical elasticity studies linearly elastic, isotropic material bodies; its goal is to find the displacement field induced in such bodies by the applied loads and the imposed confinement conditions; such goal is achieved by solving the Navier equations, in the form they were given by Cauchy in 1828.

In the problems we deal with—the first and most important of which bears the name of Lord Kelvin, who wrote about it in 1848—*a point or a line load induces a singular stress field in an unbounded body* occupying the whole of a N-dimensional space (with $N = 3$ or 2) or half of it; the strength of the singularity is $\rho^{-(N-1)}$, for ρ the distance from the point where the load is applied.

My attention for this problem class is fairly recent: it dates from an accidental inspection of the solution of one of them (Flamant's, in its two-dimensional formulation), from which I managed to dig out a first example of *concentrated contact interaction*. Since the occurrence of concentrated contact interactions was not contemplated in continuum mechanics, I felt like a naturalist who steps into a new species. In search of more examples, I turned to all problems in the class and, to push myself through the abundant relative literature, I made them the subject of an advanced undergraduate course in elasticity for students in civil engineering that I imparted in the fall of 2005. From my reading and teaching, I gradually became persuaded that all those problems could be solved in a simple, systematic, and constructive, new manner. The notes I prepared formed the first draft of this book, which I begun to refer to by its present facetious title because its contents, that the subtitle details, do cover the body of knowledge of linearly elastic geomechanics, and more.

Antonino Favata, my co-author, was one of the students sitting in my 2005 course; he showed an unusual interest in the subject and worked out some nontrivial exercises. Six years later, when I resumed my notes for a second delivery of that course, Favata had just defended his Ph.D. thesis, and held a post-doctoral position in my department. He not only helped me in many ways all along the course but also produced on his own a substantial amount of new material—chiefly, in theme of the Kelvin, Melan, and Mindlin, problems—that has been incorporated where appropriate.

Rome, October 2012 Paolo Podio-Guidugli

Preface

The problems we are going to study are all named: those after Kelvin, Boussinesq, and Cerruti, all three-dimensional, were solved explicitly during the second half of the nineteenth century (Cerruti's paper [7] appeared in 1882, long after the 1848 paper by Kelvin [18], the papers by Boussinesq [3, 4, 5, 6] between 1878 and 1892; in the same year 1892, a use of Boussinesq's solution in the guise of a Green kernel allowed for the solution of the Flamant Problem [9]). Two other problems in the same class were solved in the twentieth century: Melan's in 1932 [13]; Mindlin's, the three-dimensional version of Melan's, in 1936 and 1953 [14, 15].

All those problems were first dealt with by trying to solve the Navier equations for a displacement field having the representation constitutively implied by a tentative representation of the stress field in terms of a scalar potential. Such displacement field, in the absence of diffused volume loads, was the one induced by a *concentrated force* (or a *doublet*; see [14], Sect. 44 and 51–53; see also [4], Sect. 13.12).

When applied at an inner point, as in Kelvin's problem of a body occupying the whole space, a concentrated force was regarded as a special type of distance force. In modern terms, the Kelvin Problem can be rephrased as the problem of finding the Green function for the Navier equations in the whole space. A similar rephrasing fits both the Mindlin Problem [14, 15] and the Melan Problem [13], where a concentrated force is applied at an inner point of a half-space having dimension 3 and 2, respectively. Boussinesq, Cerruti, and Flamant considered instead concentrated forces applied to half-spaces, that is to say, they solved a boundary-value problem with a special assignment of tractions on the accessible part of the boundary. All these authors could not help to regard a concentrated force as an approximation: in the words of Love, the general problem under study was "[t]he problem… of the transmission into a solid body of force applied locally to a small part of its surface"; in short, "the problem of transmission of force" (see Love's "Historical Introduction", pp. 15 and 16 of [11]).

The original papers we quoted make an interesting and instructive reading, but generally not an easy one. Accounts at various levels of completeness and clarity are found in many textbooks, among which we mention, in addition to the quoted treatise by Love [11], those by Sokolnikoff [17], Malvern [12], Gladwell [10], Benvenuto [2], Davis and Selvadurai [8] (whose title inspired ours), Barber [1], and Sadd [16].

We propose a different approach where the basic equations of an elasticity problem—namely, balance, constitutive, and compatibility equations—are not combined into one partial differential equation for displacement, as Navier did, but instead used sequentially, in a heuristic fashion that is guided by the symmetries in the stress and displacement fields intrinsic to the problem at hand and justified a *posteriori* by appealing to the uniqueness theorem of classical linear elasticity.

Symmetries are the qualitative features of a problem that an educated intuition detects. When given an appropriate mathematical form, they dictate the choice of a priori representations for the unknown fields; such representations simplify substantially the solution process: think, for example, of the simplifications in solving St. Venant's problem induced by the a priori representation for the stress field he chose. In mechanics, the intuitive symmetries are those that the traction and displacement fields have, in duality, as a consequence of symmetries in the geometric, constitutive, and load, data; needless to say, symmetries in tractions entrain symmetries in stresses, while symmetries in strains are entrained by symmetries in displacements.

In this book, our discussion of each problem begins precisely with a careful examination of the prevailing symmetries. Here is the procedure we follow: we first look for a *general solution in terms* of stress of the balance equation; as a rule, what we find is a parametric family of solutions, among which we choose the only one turning out to be *geometrically compatible* when a linearly elastic and isotropic strain response to stress is postulated; accordingly, our choice is made by selecting the parameters so as to satisfy the *compatibility equation written in terms of stresses*. Next, we find the strain field from the stress field by a direct use of the constitutive equation; and, finally, we construct the displacement field by an explicit integration of the strain field.

It is important to realize that the strain and displacement fields we arrive at are those the applied loads induce in material bodies whose response to stress is modeled as it was almost invariably done in the nineteenth century, the golden age of classical elasticity. Now, the way most materials of geotechnical interest behave is far from being linearly elastic. Yet, it so happens that some, if not all, of the information embodied in the balanced stress states we determine do not depend on the material response, and hence play a direct and central role in designing, say, structure/foundation interactions.

In Chap. 1, our general plan of action is exemplified in an elementary one-dimensional case, where technical difficulties are minimal and yet the main conceptual ingredients are preserved. In higher dimensions, certain technical features are encountered that are different depending on whether a three- or two-dimensional version of the same problem is dealt with. This prompted us to begin by a preparatory Part I, consisting of two chapters. In Chap. 2, the basics of linear elasticity are reviewed, with special attention to the key features of classical plane elasticity. In Chap. 3, a quick account is given of curvilinear coordinates—a geometric tool that, when properly used, enhances the advantages of a preliminary inspection of symmetries—and of how differential operators are represented in non-Cartesian bases.

Part II, the bulk of this book, consists of three chapters, dedicated to, the Flamant, Boussinesq, and Kelvin Problems, respectively. In Chap. 4, the Flamant Problem is first dealt with in its simpler two-dimensional formulation; among other things, this leads to consideration of the Airy stress function. It is in this chapter that we individuate a number of significant examples of *concentrated contact interactions*, and we show how to give them an unambiguous mechanical status via the finite power expenditure they entrain. Chapter 5 begins with a discussion of a set of symmetry requirements for Boussinesq Problem that, although at first sight plausible, turn out to be geometrically inappropriate. It is then shown that, once the correct symmetries are stipulated, not only the solution is found, but also a number of related problems can be quickly solved by exploiting the superposition principle of linear elasticity, that is to say, by recognizing that the Boussinesq solution can serve as the Green function for those problems. In Chap. 6, we first show that solving Kelvin Problem by juxtaposition with seamless suture of two anti-mirror symmetric Boussinesq problems is possible only when the material is deemed incompressible; then, the two- and three-dimensional Kelvin problems are solved for compressible materials. In Chap. 7, the first of Part III, we tackle Melan's and Mindlin's problems (the former being the two-dimensional version of the latter) by the method of superposition and restriction introduced in Sect. 7.4 for their common one-dimensional version. Finally, in Chap. 8, we deal with a problem, Cerruti's, that has different symmetries and hence must be solved afresh.

Support information for various parts of the book are found in the final Appendix. In our intention, the matters surveyed there and in Part I can somehow clean up and complement the bag of knowledge of the audience we target, that is, *advanced undergraduate and graduate students in engineering and applied mathematics*. In consideration of the thorough discussion of physical motivations, the detailed presentation of heuristic arguments, and the unabridged treatment of the mathematical developments, we are convinced that teaching out of our book would be easy and rewarding. It seems to us that the book, although slim, is fairly well self-contained: the only prerequisites are a reasonable familiarity with linear algebra (in particular, manipulation of vectors and tensors) and with the usual differential operators of mathematical physics (gradient, divergence, curl, and laplacian); the few nonstandard notions are introduced with care.

Finally, to our knowledge, an equally exhaustive, compact, and consequential exposition of the classical problems listed in the subtitle is not found anywhere else. Thus, we hope our booklet will also serve as a reference for students in elasticity.

References

1. Barber JR (2002) Elasticity. Kluwer, Dordrecht
2. Benvenuto E (1981) La Scienza delle Costruzioni e il suo Sviluppo Storico. Sansoni, Firenze
3. Boussinesq J (1878) Équilibre d'élasticité d'un sol isotrope sans pesanteur, supportant differents poids. CR Acad Sci 86:1260–1263
4. Boussinesq J (1885) Application des Potentiels à l'Étude de l'Équilibre et du Mouvement des Solides Élastiques. Gauthiers-Villars, Paris
5. Boussinesq J (1888), Équilibre d'élasticité d'un solide sans pesanteur, homogéne et isotrope, dont les parties profondes sont maintenues fixes, pendant que sa surface éprouve des pressions ou des déplacements connus, s'annullant hors d'une région restreinte oú ils sont arbitraires. CR Acad Sci 106:1043–1048, 1119–1123
6. Boussinesq J (1892) Des perturbations locales que produit au-dessous d'elle une forte charge, répartie uniformément le long d'une droite normale aux deux bords, à la surface supérieure d'une poutre rectangulaire et de longueur indéfinie posée de champ soit sur un sol horizontal, soit sur deux appuis transversaux équidistants de la charge. CR Acad Sci 114:1510–1516
7. Cerruti V (1882) Ricerche intorno all'equilibrio de' corpi elastici isotropi. Rend Accad Lincei 3(13):81–122
8. Davis RO, Selvadurai APS (1996) Elasticity and geomechanics. Cambridge University Press, Cambridge
9. Flamant A (1892) Sur la répartition des pressions dans un solide rectangulaire chargé transversalement. CR Acad Sci 114:1465–1468
10. Gladwell GML (1980) Contact problems in the classical theory of elasticity. Sijthoff and Noordhoff, Alphen aan den Rijn
11. Love AEH (1927) A treatise on the mathematical theory of elasticity. Cambridge University Press, Cambridge
12. Malvern LE (1969) Introduction to the Mechanics of a Continuous Medium. Prentice-Hall, Englewood Cliffs
13. Melan E (1932) Der Spannungszustand der durch eine Einzelkraft im Innern beanspruchten Halbscheibe. Z Angew Math Mech 12:343–346
14. Mindlin RD (1936) Force at a point in the interior of a semi-infinite solid. Physics 7:195–202
15. Mindlin RD (1950) Nuclei of strain in the semi-infinite solid. J Appl Phys 21:926–930
16. Sadd MH (2009) Elasticity: Theory, applications, and numerics. Elsevier Butterworth Heinemann, Amsterdam
17. Sokolnikoff IS (1956) Mathematical theory of elasticity. McGraw-Hill, New York
18. Thompson W (Lord Kelvin) (1848) Note on the integration of the equations of equilibrium of an elastic solid. Cambr Dubl Math J 3:87–89

Contents

1 One-Dimensional Paradigms 1
- 1.1 Integration Methods, Standard and Not 1
- 1.2 Green-Kernel Integration 4
- 1.3 The One-Dimensional Kelvin Problem 7
- 1.4 The One-Dimensional Mindlin Problem 10

Part I Preliminaries

2 Elements of Linear Elasticity 17
- 2.1 Displacement, Strain, Compatibility 17
  - 2.1.1 Compatibility 22
  - 2.1.2 Plane Displacement Fields 23
  - 2.1.3 Plane Strain Fields 24
- 2.2 Forces, Stress, Equilibrium 26
  - 2.2.1 Cauchy's Notion of Stress 27
  - 2.2.2 Free-Body Diagrams, Diffuse and Concentrated Forces 28
- 2.3 The Stress Response to Strain 30
  - 2.3.1 Isotropic Materials 31
  - 2.3.2 Mechanical Interpretation of the Elastic Moduli 32
  - 2.3.3 Plane Stress Fields 36
  - 2.3.4 Plane Strain Fields Associated with Plane Stress Fields 37
- 2.4 Compatibility in Stress Terms 38
  - 2.4.1 The Three-Dimensional Case 38
  - 2.4.2 The Two-Dimensional Case 39
- References 40

3 Geometric and Analytic Tools 43
- 3.1 Curvilinear Coordinates, Covariant and Contravariant Bases 43

3.2 Orthogonal Coordinates, Physical Bases 44
3.2.1 Polar Coordinates 44
3.2.2 Cylindrical Coordinates 46
3.2.3 Spherical Coordinates 48
3.3 Representations of Differential Operators 51

Part II Three Classical Problems: Flamant's, Boussinesq's, and Kelvin's

4 The Flamant Problem 55
4.1 A Priori Representations of Displacement, Strain, and Stress . . . 55
4.2 Plane Stress Fields Balanced for Null Distance Forces. 58
4.3 The 2-D Boussinesq–Flamant Problem. 60
4.3.1 Divergenceless Plane Stress Fields 60
4.3.2 Compatible and Divergenceless Plane Stress Fields. 62
4.3.3 Compatible and Divergenceless Plane Stress Fields that Balance the Applied Force 63
4.4 Digression 65
4.4.1 A Weak Formulation of the Boussinesq–Flamant Problem. 65
4.4.2 Concentrated Contact Interactions. 67
4.5 The Flamant Elastic State. 70
4.5.1 The Stress Field 70
4.5.2 The Strain Field 70
4.5.3 The Displacement Field. 72
References 77

5 The Boussinesq Problem 79
5.1 Preliminary Symmetry Considerations 79
5.2 Inexistence of Center-Symmentric Solutions. 81
5.2.1 Balanced Center-Symmetric Stress Fields 81
5.2.2 Incompatibility of Center-Symmetric Stress Fields 83
5.3 Balanced Stress Fields with Cylindrical Symmetry 85
5.3.1 Boundary Conditions. 87
5.3.2 A Kinematic Condition 87
5.3.3 Balanced and Compatible Stress Fields 91
5.3.4 The Trace of the Stress Field 94
5.4 The Boussinesq Stress Field 99
5.4.1 Step 2 99
5.4.2 Step 3 99
5.4.3 Step 4 101

5.5 Digression . . . 102
5.5.1 Concentrated Contact Interactions . . . 102
5.5.2 How to Find Flamant'S Stress Field Having Found Boussinesq's . . . 104
5.6 The Boussinesq Strain and Displacement Fields . . . 104
5.7 Exploiting the Boussinesq Solution . . . 108
5.7.1 The Pressure Bulb . . . 108
5.7.2 Displacements in a Half-Space Subject to a Diffused Load . . . 109
5.7.3 Displacements in a Half-Space Subject to a Diffused Load . . . 111
5.7.4 Horizontal Surface Displacements in the Flamant Problem . . . 112
5.8 An Alternative Representation of the Boussinesq Elastic State . . . 113
References . . . 114

6 The Kelvin Problem . . . 115
6.1 Solution by Juxtaposition . . . 115
6.1.1 Continuity Conditions at Sutures . . . 115
6.1.2 Conditional Solvability: The Boussinesq Problem for Incompressible Materials . . . 117
6.2 The 2-D Kelvin Problem . . . 120
6.2.1 Balanced Stress Fields . . . 121
6.2.2 Compatible Stress Fields . . . 123
6.2.3 Strain and Displacements Fields . . . 123
6.3 The Kelvin Elastic State . . . 126
6.3.1 The Stress Field . . . 126
6.3.2 The Strain and Displacement Fields . . . 128
References . . . 129

Part III Three Other Problems: Melan's, Mindlin's, and Cerruti's

7 The Melan and Mindlin Problems . . . 133
7.1 Solution by Superposition . . . 133
7.2 The Melan Problem . . . 134
7.2.1 The Stress Field . . . 134
7.2.2 The Strain and Displacement Fields . . . 139
7.3 The Mindlin Problem . . . 140
7.3.1 The Stress Field . . . 141
7.3.2 The Strain and Displacement Fields . . . 146
References . . . 148

8 The Cerruti Problem 149
8.1 Displacement and Stress Symmetries 149
8.2 The Stress Field of the 2-D Cerruti Problem 151
8.3 The Diffused-Load Cerruti Problem 154
8.3.1 The Stress and Strain Fields 154
8.3.2 The Displacement Field 155
References 157

Postface 159

Appendix A 161

Index 177

Symbols

e_{ijk}	Ricci's symbol
f	Concentrated load on a beam
q	Diffused load on a beam
w	Axial displacement of a beam
o	Space point chosen as coordinate origin
x	Typical space point
x_i	Cartesian coordinates
$\boldsymbol{c}$	Contact actions
$\boldsymbol{d}$	Distance actions
$\boldsymbol{d}^{in}$	Inertial distance actions
$\boldsymbol{d}^{ni}$	Non-inertial distance actions
$\boldsymbol{e}$	Unit vector, radial in polar coordinates
$\boldsymbol{e}_i$	Cartesian basis of $\mathcal{V}$
$\boldsymbol{g}_i$	Covariant basis of $\mathcal{V}$
$\boldsymbol{g}^i$	Contravariant basis of $\mathcal{V}$
$\boldsymbol{g}_{<i>}$	Physical basis of $\mathcal{V}$
$\boldsymbol{n}$	Outer normal vector
$\boldsymbol{u}$	Displacement vector
$\boldsymbol{x}$	Position vector of x with respect to o
A	Area of the cross section of a beam
E	Young modulus
E_0	Plane Young modulus
G	Shear modulus
$G(\cdot,\cdot)$	Green function
H	Heaviside function
L	Differential operator
N	Normal force in a beam
R	Region of $\mathcal{E}^N$
$\boldsymbol{E}$	Strain tensor
$\boldsymbol{I}$	Identity tensor
$\boldsymbol{I}_{(2)}$	Two-dimensional identity tensor

$\boldsymbol{S}$	Cauchy stress tensor
$\boldsymbol{S}^{(A)}$	Active stress tensor
$\boldsymbol{S}^{(R)}$	Reactive stress tensor
$\boldsymbol{W}$	Rotation tensor
$\mathcal{B}_\rho$	Ball of radius ρ
$\mathcal{D}_\rho$	Disk-shaped region of radius ρ
$\mathcal{E}^N$	N-dimensional Euclidean space
$\mathcal{HC}_\rho$	Half-circumference of radius ρ
$\mathcal{HD}_\rho$	Half-disk-shaped region of radius ρ
$\mathcal{HP}^\pm$	Half-planes
$\mathcal{HS}^\pm$	Half-spaces
$\mathcal{QD}_\rho$	Quarter-disk-shaped region of radius ρ
$\mathcal{R}$	Real line
$\mathcal{V}$	Vector space associated with $\mathcal{E}^N$
$\mathbb{C}$	Elasticity tensor
$\mathbb{C}$	(in Section A.7) Set of complex numbers
$\mathbb{R}$	Set of real numbers
δ	Dirac delta
δ_{ij}	Kronecker symbol
ζ^i	Curvilinear coordinates
λ,μ	Lamé moduli
ν	Poisson modulus
ν_0	Plane Poisson modulus
σ	Strain energy per unit volume
curl	Curl, of a vector or tensor field
det	Determinant
dim	Dimension, of a vector space or a physical quantity
div	Divergence, of a vector or tensor field
sgn	Sign function
tr	Trace operator
$(\cdot)^T$	Transposition operator
Lin	Space of second-order tensors
Skw	Space of skew-symmetric tensors
Sym	Space of symmetric tensors
$[\![\cdot]\!]$	Jump operator
$\cdot$	Scalar product
$\times$	Vector product
$\otimes$	Dyadic product
:=	Definition symbol

Chapter 1
One-Dimensional Paradigms

In this introductory chapter, we work in a one-dimensional (1-D) setting. Firstly, we exemplify the nonstandard integration method we are going to use systematically in Part II. Secondly, we exemplify the Green-kernel integration method to be exploited, in particular, for the problems collected in Part III. Finally, we use these two integration methods to solve the 1-D versions of Kelvin's and Mindlin's problems. We invite the reader to return to this chapter after studying the developments in Part II and III, so as to experience the subtle pleasure of looking at a mathematically elementary subject from a superior point of view.

Our notations are those widely used in engineering mechanics, but we make a systematic effort to suggest, through the use of the standard terminology of continuum mechanics, the kinship between the 1-D objects we here manipulate and their 3-D counterparts to be introduced in the next chapter.

1.1 Integration Methods, Standard and Not

Consider a linearly elastic straight beam, of length l, cross-section area A, and Young modulus E, whose axial stiffness EA may depend on the axial coordinate z: see Fig. 1.1, where both ends are shown hinged, a choice irrelevant to the substance of our developments to come. The applied axial load q (a *distance force*, in the terminology of continuum mechanics) is diffused; it induces an internal stress measured by the normal force N, the 1-D counterpart of the *stress tensor* $\mathbf{S}$. The axial deformation is denoted by ε, the axial displacement by w; their 3-D counterparts are the *strain tensor* $\mathbf{E}$ and the *displacement vector* $\boldsymbol{u}$, respectively.

The beam's equilibrium problem is ruled by the following field equations:

$$N' + q = 0, \qquad (\textit{balance}\text{ eq.}) \tag{1.1}$$

$$\varepsilon = w', \qquad (\textit{compatibility}\text{ eq.}) \tag{1.2}$$

$$N = (EA)\varepsilon, \qquad (\textit{constitutive}\text{ eq.}) \tag{1.3}$$

P. Podio-Guidugli and A. Favata, *Elasticity for Geotechnicians*, Solid Mechanics and Its Applications 204, DOI: 10.1007/978-3-319-01258-2_1,

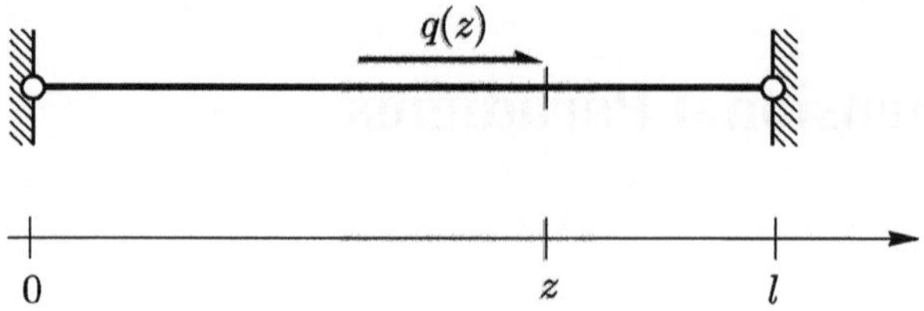

Fig. 1.1 A doubly-hinged beam subject to a diffused axial load

(a prime signifies differentiation); the following boundary equations reflect the presence of the end hinges:

$$w(0) = 0 = w(l), \qquad (\textit{boundary}\ \text{eq.s}) \tag{1.4}$$

According to the standard method of solution, the field equations are combined so as to arrive at a single equation for w; since this equation is differential, one speaks of an *integration method.* Here is the flow chart:
firstly, the constitutive and compatibility equations are combined;

$$N = (EA)w'; \tag{1.5}$$

secondly, this last relation is substituted in the balance equation, yielding:

$$((EA)w')' + q = 0; \tag{1.6}$$

two successive integrations introduce two constants, that are determined by the use of the boundary conditions.

The integration method we are going to use is different, in that it produces a solution by taking into account the equations of system (1.1)–(1.4) in a different order.

The first step consists in integrating the balance Eq. (1.1), whose general solution:

$$N(z) = N(0) - \int_0^z q(s)ds \tag{1.7}$$

depends on the parameter $N(0)$. Now, a normal force field is *compatible* if, by making use of the inverse of the constitutive relation (1.3):

$$\varepsilon = (EA)^{-1}N,$$

it can be associated with an axial deformation field from which a displacement field w can be constructed according to the compatibility Eq. (1.2); that is to say, a normal force field is compatible if it induces a displacement field of the following form:

$$w(z) = w(0) + \int_0^z \varepsilon(\zeta)d\zeta = w(0) + \int_0^z \frac{N}{EA}(\zeta)d\zeta\,. \tag{1.8}$$

Hence, the minimal compatibility requirement for a field N having the representation (1.7) is that the function

$$\zeta \mapsto \frac{N}{EA}(\zeta)$$

be integrable over the interval $(0, l)$; this requirement is easy to guarantee, whenever the axial stiffness is reasonably smooth, given that relation (1.7) implies the continuity of the normal force.

Once the needed smoothness is ascertained, the second step consists simply in combining (1.8) with (1.7), so as to arrive at

$$w(z) = w(0) + \int_0^z \frac{1}{EA(\zeta)}\Big(N(0) - \int_0^\zeta q(s)ds\Big)d\zeta;$$

finally, the imposition of the boundary conditions (1.4) determines uniquely the values of parameters $w(0)$ and $N(0)$.

Remark 1.1 The normal force field corresponding to the solution of problem (1.1)–(1.4) is obtained by substituting into (1.7) the value of $N(0)$ satisfying the condition that both ends of the beam under study do not move, namely,

$$0 = \Big(\int_0^l \frac{1}{EA(\zeta)}d\zeta\Big)N(0) - \int_0^l \Big(\frac{1}{EA(\zeta)}\int_0^\zeta q(s)ds\Big)d\zeta.$$

For general assignments of the data, this is the best we can do. But, it is not rare that information of this sort can be read off from the posing of the problem at hand: for example, if both axial stiffness and load are constant-valued, then it is easy to see, by symmetry, that

$$N(0) = \frac{1}{2}ql.$$

Throughout this book, to ease the solution process of all 2- and 3-D equilibrium problems we deal with, we shall be making a systematic use of symmetries in the data assignment.

Remark 1.2 The success of our integration method hinged on the invertibility of the constitutive equation (which is induced by assuming that $EA(z) > 0$ for $z \in [0, l]$, so as to guarantee the model's physical plausibility) and on the assumed smoothness of the data (both functions q and $(EA)^{-1}$ must be integrable over the interval $(0, 1)$). Such conditions have direct counterparts in the case of elasticity problems in dimension greater than 1. However, when we exploit our integration method for anyone of those problem, it so happens that the compatibility requirement for an equilibrium stress field $\mathbf{S}$ does not consist only in asking that it is conveniently smooth, but also that it satisfies a second-order partial differential equation. A 1-D context is too poor to keep track of all the complications we are going to face in dimension 2 or 3. Similarly, in such a context there is a scarce or null chance of illustrating in an adequate manner another characteristic trait of our method, that is, the just mentioned preliminary examination of intrinsic symmetries that often allows for an explicit parametric representation of the unknown fields.

Remark 1.3 Formulation (1.6) is the so-called *strong formulation* of the problem we considered. With such a formulation, very regular solutions of the problem are sought: e.g., in the case of the boundary-value problem depicted in Fig. 1.1, we seek $w \in C^2(]0,l[) \cap C([0,l])$; this condition becomes even stricter, namely, $w \in C^2(]0,l[) \cap C^1(]0,l])$, if the hinge at the right end is replaced by a simple support where a force P in the axial direction is applied, because in this case the homogeneous *Dirichlet*-type boundary condition $w(l) = 0$ is replaced by the inhomogeneous *Neumann*-type boundary condition $N(l) = P$, or rather, in view of (1.5), by $(EA\, w')(l) = P$.

1.2 Green-Kernel Integration

Let us consider again the beam we studied in Sect. 1.1, this time subject to an axial force $\boldsymbol{f} = f\boldsymbol{e}$ applied in the section of abscissa z_0, with $0 < z_0 < l$ (Fig. 1.2), and assume that the axial stiffness *EA* is constant. Our purpose is to find an analytic expression for the axial displacement $w = w(z, z_0; f)$.

To this end, we observe that the boundary reactions ($-af$ at $z = 0$ and $-bf$ at $z = l$, with both a and b positive) are determined by the balance equation:

$$a + b = f \tag{1.9}$$

and the continuity condition at $z = z_0$ for the displacement:

$$a\, z_0 = b(l - z_0); \tag{1.10}$$

the solution of system (1.9)–(1.10) is:

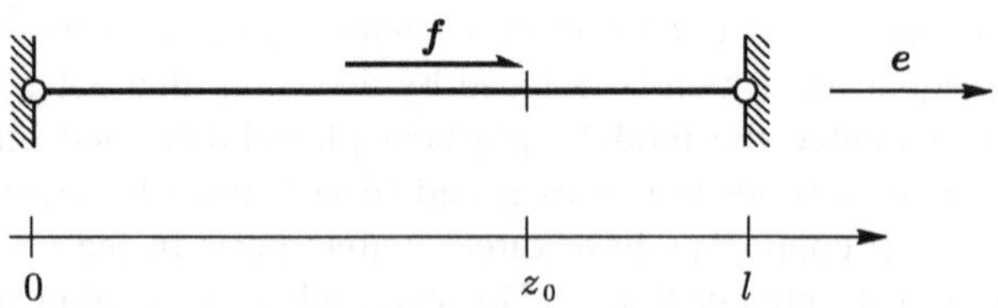

Fig. 1.2 A doubly-hinged beam subject to a concentrated axial load

$$a = f\left(1 - \frac{z_0}{l}\right), \quad b = f\frac{z_0}{l}.$$

The left part of the beam, where the strain is $(EA)^{-1}a$, lengthens, by $(EA)^{-1}az_0$ units; the right part shortens, by $(EA)^{-1}b(l - z_0)$ units; the total length cannot change. Consequently, the normal force is piece-wise constant:

$$\begin{aligned} N(z) &= f\left(1 - \frac{z_0}{l}\right) && \text{for } 0 \leq z < z_0, \\ &= -f\frac{z_0}{l} && \text{for } z_0 < z \leq l, \end{aligned}$$

and the same is true for the axial strain

$$\varepsilon = \frac{N}{EA}.$$

Hence, to satisfy the compatibility condition

$$\varepsilon = w',$$

we find that

$$\begin{aligned} w(z, z_0; f) &= \frac{f}{EA}\left(1 - \frac{z_0}{l}\right)z && \text{for } 0 \leq z < z_0, \\ &= \frac{f}{EA}\left(1 - \frac{z}{l}\right)z_0 && \text{for } z_0 \leq z \leq l. \end{aligned} \tag{1.11}$$

Remark 1.4 Note that

$$\frac{f}{EA}\left(1 - \frac{z}{l}\right)z_0 = \frac{f}{EA}\left(1 - \frac{z_0}{l}\right)z_0 - \frac{f}{EA}\frac{z_0}{l}(z - z_0),$$

an expression in terms of (1.11) of the identity of

$$w(z, z_0; f) = w(z_0, z_0; f) - \big(w(z, z_0; f) - w(z_0, z_0; f)\big).$$

Note also that

$$w(z, z_0; f) = w(z_0, z; f), \tag{1.12}$$

i.e., that the displacement induced at z by a force applied at z_0 equals the displacement induced at the latter point when the same force is applied in the former, an elementary manifestation of a basic result in 3-D linear elasticity, the *Reciprocity Theorem* of Enrico Betti (1823–1892).

Let us now rewrite the expression (1.11) for the axial displacement in a compact form:

$$w(z, z_0; f) = \frac{f}{EA}\Big(\Big(1 - \frac{z_0}{l}\Big)z - H(z - z_0)(z - z_0)\Big), \quad 0 \le z \le l,$$

where H is the restriction to the interval $[0, l]$ of *Heaviside's step function* (Sect. A.1):

$$\begin{aligned} H(z - z_0) &= 0 \quad \text{per } 0 \le z \le z_0, \\ &= 1 \quad \text{per } z_0 < z \le l. \end{aligned}$$

By definition, the *displacement Green function* for the boundary-value problem under examination is:

$$G_w(z, z_0) := w(z, z_0; 1) = \frac{1}{EA}\Big(\Big(1 - \frac{z_0}{l}\Big)z - H(z - z_0)(z - z_0)\Big). \tag{1.13}$$

This function can be regarded as the solution of the linear differential problem

$$L[w] = \delta(z, z_0)\, f, \quad L[w] := -(EA)w'',$$

where δ is the *Dirac delta* (Sect. A.1), when the boundary conditions are:

$$w(0) = 0 = w(l). \tag{1.14}$$

If the beam is subject to a diffuse axial load q, the corresponding displacement field is given by the formula:

$$w(z; q) = \int_0^l G_w(z, \zeta)q(\zeta)\, d\zeta.$$

We can interpret formally this result as an implicit definition of the *linear integral operator* L^{-1} that inverts the *linear differential operator* L under the boundary conditions (1.14); and we can write:

$$w = L^{-1}[q] = \int_0^l G_w(z, \zeta)q(\zeta)\, d\zeta, \tag{1.15}$$

where the Green function is the so-called *kernel* of the integral representation for L^{-1}. Moreover, the normal force field due to the diffuse load q can be written as follows:

$$N(z) = \int_0^l G_N(z, \zeta)q(\zeta)\, d\zeta, \quad G_N(z, \zeta) = 1 - \frac{\zeta}{l} - H(z, \zeta),$$

where the *stress Green function* G_N is the kernel of the integral operator that inverts formally the equilibrium differential operator

$$\widetilde{L}[N] := -N'.$$

Remark 1.5 The displacement Green function is symmetric:

$$G_w(z_1, z_2) = G_w(z_2, z_1),$$

because the operator L is *self-adjoint*, that is, satisfies the following condition:

$$< u, L[v] > := \int_0^l u(\zeta) L[v(\zeta)]\, d\zeta = \int_0^l v(\zeta) L[u(\zeta)]\, d\zeta = < v, L[u] >,$$

for all pairs of regular fields u, v defined over $[0, l]$ that obey the prescribed boundary conditions.

Self-adjointness of operator L is the crucial property for the validity of Betti's Reciprocity Theorem. In fact, let the beam in Fig. 1.2 be loaded by a unit force applied at z_2 first, then at z_1, with $z_1 < z_2$. With a use of Green representation (1.15) for the inverse operator, we find:

$$w(z_1, z_2; 1) = \int_0^l G_w(z_1, \zeta)\delta(z_2, \zeta)\, d\zeta = G_w(z_1, z_2) = \left(1 - \frac{z_2}{l}\right) z_1,$$

$$w(z_2, z_1; 1) = \int_0^l G_w(z_2, \zeta)\delta(z_1, \zeta)\, d\zeta = G_w(z_2, z_1) = \left(1 - \frac{z_2}{l}\right) z_1$$

(cf. (1.12) and (1.13)).

1.3 The One-Dimensional Kelvin Problem

In this section, we deal with a problem which can be regarded as the 1-D version of the Kelvin problem to be tackled in Chap. 5, i.e., the equilibrium problem of an infinite linearly elastic rod, subject to an axial load applied at a point, which we identify with the origin of the coordinates (Fig. 1.3).

Our intention is to solve this problem by the same sequence of operations we shall use for its 2- and 3-D versions.

In order to underline this similitude as much as possible, we give the elastic state in the beam a redundant representation, beginning with the normal force, that we represent in tensorial form as a dyad[1]:

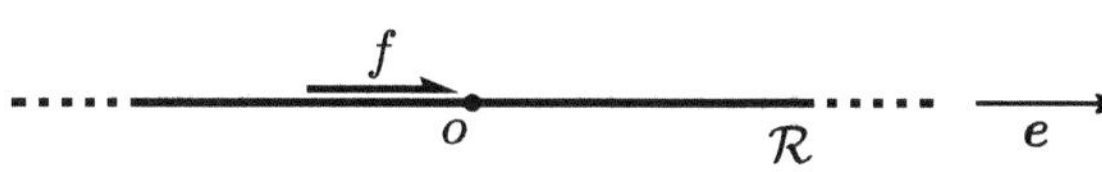

Fig. 1.3 An infinite beam subject to a concentrated axial load

---

[1] The *dyadic product* of two vectors $\boldsymbol{a}, \boldsymbol{b}$ is defined as follows:

$$(\boldsymbol{a} \otimes \boldsymbol{b})\boldsymbol{v} := (\boldsymbol{b} \cdot \boldsymbol{v})\boldsymbol{a}, \quad \text{for all vectors } \boldsymbol{v}.$$

$$\boldsymbol{N}(z) = N(z)\boldsymbol{e} \otimes \boldsymbol{e}\,.$$

We require the field $\boldsymbol{N}$ to satisfy, *in the sense of distributions on the real line* $\mathcal{R}$, the differential equation:

$$\operatorname{div} \boldsymbol{N} + \boldsymbol{f} = \boldsymbol{0}, \quad \boldsymbol{f} = f\delta(0)\boldsymbol{e}, \tag{1.16}$$

an equation that can be rewritten in the form:

$$\big(N' + f\delta(0)\big)\boldsymbol{e} = \boldsymbol{0},$$

where the divergence operator is reduced to the differentiation operator. For $\boldsymbol{v} = v(z)\boldsymbol{e}$ a test function with compact support, Eq. (1.16) assumes a more meaningful form:

$$\int_{\mathcal{R}} \big(N' + f\delta(0)\big)\boldsymbol{e} \cdot \boldsymbol{v}\, dz = \int_{\mathcal{R}} \big(N' + f\delta(0)\big) v\, dz = 0,$$

or rather, and better, the distributional form:

$$\int_{\mathcal{R}} N v' dz = f\, v(0) \quad \text{if } 0 \in \operatorname{supp}(v), \qquad N' = 0 \quad \text{otherwise.} \tag{1.17}$$

On denoting with $N(0\pm)$ the values that the piece-wise constant function $z \mapsto N(z)$ assumes on the half-lines $\mathcal{R}^{\pm}$, the first of equations (1.17) can be written in the final form:

$$[\![N]\!]_0 + f = 0, \tag{1.18}$$

where the divergence operator in (1.16) is replaced by the *jump operator*:

$$[\![N]\!]_z := N(z+) - N(z-), \tag{1.19}$$

evaluated at $z = 0$.[2]

In conclusion, here is how the normal force field looks like:

$$\boldsymbol{N}(z) = -\frac{1}{2} f \operatorname{sgn}(z)\, \boldsymbol{e} \otimes \boldsymbol{e}, \quad z \neq 0 \tag{1.20}$$

(the sgn function is defined in Sect. A.1.) the relative Green function is:

$$G_N(z, \zeta) = -\frac{1}{2} \operatorname{sgn}(z - \zeta)\,.$$

---

[2] To obtain (1.18), it is sufficient to note that

$$\int_{\mathcal{R}} N v' dz = N(0-) \int_{\mathcal{R}^-} v' dz + N(0+) \int_{\mathcal{R}^+} v' dz = -v(0)\big(N(0+) - N(0-)\big),$$

The normal-force field plays here the role of Kelvin's 3-D stress field. It is interesting to see that a concentrated load induces at the point where it is applied a *discontinuous* normal force in dimension 1, and a *singular* stress in dimension 2 or 3.

We complete the determination of the elastic state in the beam by oberving that the inverse constitutive equation yields:

$$\boldsymbol{E}(z) = (EA)^{-1}\boldsymbol{N}(z);$$

the displacement field $\boldsymbol{u} = u\,\boldsymbol{e}$ can be found by integration of the differential equation:

$$\left(\boldsymbol{u}' \otimes \boldsymbol{e} = \boldsymbol{E} \quad \Leftrightarrow\right) \; u' = -\frac{f}{2\,EA}\,\mathrm{sgn}\,(z), \quad z \neq 0,$$

with the transmission condition $[\![u]\!]_0 = 0$. Integration yields:

$$u(z) = u(0) - \frac{f}{2\,EA}\,|z|\,; \tag{1.21}$$

$u(0)$ denotes a rigid translation which remains arbitrary, since no kinematic constraint is available to determine it.

Remark 1.6 Solving the Kelvin problem is important not only in itself but also because it allows to construct by differentiation or integration an arbitrary number of *strain nuclei*, that is to say, of singular solutions associated with load systems consisting of more than one concentrated force. By a suitable combination of deformation nuclei, followed at times by an operation of restriction to the domain of interest, many classical problems in linear elasticity can be solved, where the applied loads are not necessarily concentrated, as for example the problem considered by Lamé of a pressurized cavity in an infinite medium.

In a 1-D context, there is only one type of deformation nucleus, to which both 3-D nuclei called *double force* and *compression/dilatation center* reduce; we now proceed to construct it.

Let us apply to the elastic beam of infinite length we have been considering in this section both a force $h^{-1} f\boldsymbol{e}$ at the point of abscissa $z = +h/2$ and the force $-h^{-1} f\boldsymbol{e}$ at $z = -h/2$. By linearity, anyone of the fields $f\,c(z,h)$ of which the elastic state of the beam consists can be evaluated at any typical point z in terms of the corresponding field $f\,c_K(z)$ obtained by solving the Kelvin problem, by means of the following formula:

$$c(z,h) = h^{-1}c_K(z - h/2) - h^{-1}c_K(z + h/2) = -\frac{c_K(z + h/2) - c_K(z - h/2)}{h},$$

whence

$$\lim_{h\to 0} c(z,h) = -c_K'(z).$$

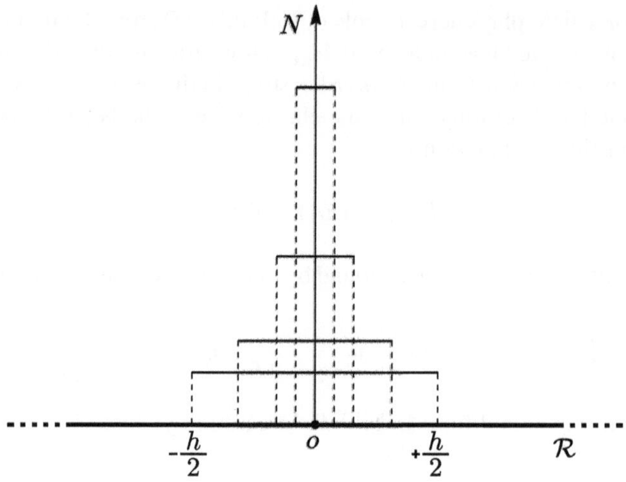

Fig. 1.4 A visualization of the graph of the mapping $h \mapsto (\text{sgn}\,(z+h/2) - \text{sgn}\,(z-h/2))/2h$ for h smaller and smaller

In case we take c_K to be the field $f^{-1}\boldsymbol{N}$ specified by (1.20), we have:

$$h^{-1}N(z-h/2) - h^{-1}N(z+h/2) = \frac{1}{2}\,\frac{\text{sgn}\,(z+h/2) - \text{sgn}\,(z-h/2)}{h} \underset{h\to 0}{\longrightarrow} \delta(z),$$

where the limit process is the well-known one suggested by Fig. 1.4. In case c_K is identified with the displacement field (1.21), we have:

$$h^{-1}u(z-h/2) - h^{-1}u(z+h/2) = \frac{1}{2\,EA}\,\frac{|z+h/2| - |z-h/2|}{h} \underset{h\to 0}{\longrightarrow} \text{sgn}\,(z)\,.$$

All in all, we say that a *dilatation center* is found at the origin when the normal force and displacement fields have the following form:

$$\boldsymbol{N}_{DC}(z) = f\,\delta(z)\,\boldsymbol{e}\otimes\boldsymbol{e}, \quad \boldsymbol{u}_{DC}(z) = f\,\text{sgn}\,z\,\boldsymbol{e}$$

(needless to say, the first of these relations must be interpreted in the sense of distributions on $\mathcal{R}$).

1.4 The One-Dimensional Mindlin Problem

In this section, we consider a problem that can be regarded as the 1-D version of the Mindlin problem (Sect. 7.3): a semi-infinite linearly elastic beam, subject to an axial load, applied at an internal point; the reader is referred to Fig. 1.5 for notations.

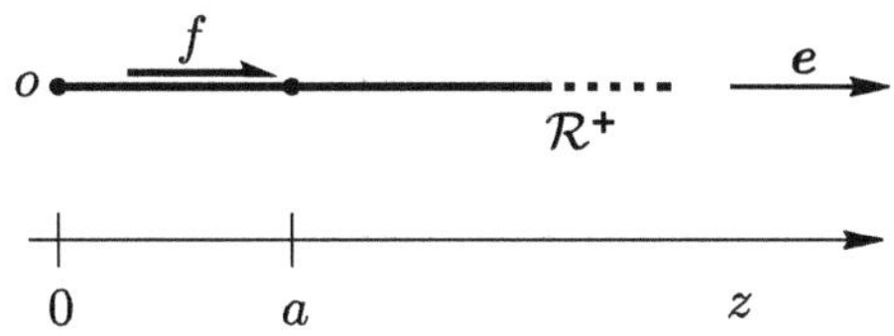

Fig. 1.5 A semi-infinite beam subject to an axial load applied at an internal point

We look for a normal force field

$$\boldsymbol{N} = N(z)\boldsymbol{e} \otimes \boldsymbol{e}$$

on the half-line $\mathcal{R}^+$, such as to satisfy the equilibrium equation

$$\big(N' + f\delta(a)\big)\boldsymbol{e} = \boldsymbol{0}$$

and the boundary condition

$$\Big(\boldsymbol{N}(0)\boldsymbol{e} = \boldsymbol{0}\Big) \quad \Leftrightarrow \quad N(0) = 0. \tag{1.22}$$

A weak formulation of this problem is: to find a piece-wise continuously differentiable function N defined over $\mathcal{R}^+$, such that, for all test functions $\boldsymbol{v} = v(z)\boldsymbol{e}$ with compact support in $\mathcal{R}^+$,

$$[\![N]\!]_a + f = 0 \ \text{ if } a \in \text{supp}(v), \qquad N' = 0 \ \text{ otherwise.}$$

It is not difficult to see that the solution is:

$$\boldsymbol{N}(z) = -f\, H(z-a)\,\boldsymbol{e} \otimes \boldsymbol{e},$$

where H is the restriction to $[0, +\infty)$ of the Heaviside function. The associated strain field is:

$$\boldsymbol{E}(z) = -\frac{f}{EA}\,H(z-a)\,\boldsymbol{e} \otimes \boldsymbol{e}.$$

With a view toward constructing the relative displacement field, we note that

$$\boldsymbol{u}'(z) = u'(z)\boldsymbol{e} = \boldsymbol{E}(z)\boldsymbol{e} = -\frac{f}{EA}\,H(z-a)\,\boldsymbol{e} \quad \Rightarrow \quad u'(z) = -\frac{f}{EA}\,H(z-a).$$

To integrate the last relation, we recall that

$$H(t) = \frac{1}{2}\left(1 + \text{sgn}\, t\right),$$

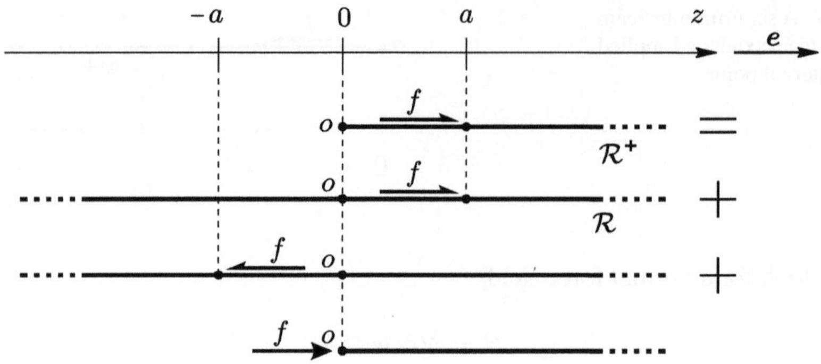

Fig. 1.6 The 1-D Mindlin's problem as a superposition of problems

and that the sign function is the distributional derivative of the modulus function. With the use of these facts, we deduce that

$$\boldsymbol{u}(z) = \Big(u(a) - \frac{f}{EA}\,\big(H(z-a)|z-a|\big)\Big)\boldsymbol{e}\,.$$

Remark 1.7 In preparation for the study of the higher-dimensional situation to be dealt with in Chap. 7, it is instructive to arrive at the solution of the 1-D Mindlin problem by superposition, with an iterated use of Kelvin solution (1.20). The procedure we propose is visualized in Fig. 1.6. Here are the steps:

(i) in a infinite beam subject to a load $f\boldsymbol{e}$ applied at $z = a$, the normal-force field is, to within an origin translation, Kelvin's:

$$\breve{\boldsymbol{N}}(z) = -\frac{1}{2}\, f\, \mathrm{sgn}\,(z-a)\,\boldsymbol{e}\otimes\boldsymbol{e}, \quad z \neq a; \tag{1.23}$$

(ii) likewise, when the load $-f\boldsymbol{e}$ is applied at $z = -a$, Kelvin's normal-force field is:

$$\widehat{\boldsymbol{N}}(z) = \frac{1}{2}\, f\, \mathrm{sgn}\,(z+a)\,\boldsymbol{e}\otimes\boldsymbol{e}, \quad z \neq -a; \tag{1.24}$$

(iii) superposing these two fields on $\mathcal{R}$ yields the field:

$$\breve{\boldsymbol{N}}(z) + \widehat{\boldsymbol{N}}(z) = -\frac{1}{2}\, f\,\big(\mathrm{sgn}\,(z-a) - \mathrm{sgn}\,(z+a)\big)\,\boldsymbol{e}\otimes\boldsymbol{e},$$

whose restriction to the half-line $\mathcal{R}^+$ is:

$$\widetilde{\boldsymbol{N}}(z) = f\big(1 - H(z-a)\big)\,\boldsymbol{e}\otimes\boldsymbol{e}, \quad z \geq 0;$$

(iv) finally, to satisfy the boundary condition (1.22), we add to $\widetilde{\boldsymbol{N}}$ the constant field

$$\overline{\boldsymbol{N}}(z) = -f\,\boldsymbol{e} \otimes \boldsymbol{e},$$

that is, the normal-force field in the lowest beam in Fig. 1.6.

We anticipate that the last step turns out to be fairly more difficult to take in the 3-D case, when the construction of an adscititious field guaranteeing fulfilment of the free-boundary condition will be performed by having recourse to a Green-kernel integration.

Part I
Preliminaries

Chapter 2
Elements of Linear Elasticity

In this chapter we give a short and yet fairly complete exposition of the elemental features of classic elasticity having relevance to our subject matters. This archetypal theory, probably the most successful and best well-known theory of continuum mechanics, has been given many excellent and exhaustive expositions. Among the textbooks including an ample coverage of the problems we deal with in this book we cite those by Love [8], Sokolnikoff [17], Malvern [9], Gladwell [5]; we also take from the Handbuch article by Gurtin [6], whose use of direct notation we find appropriate to avoid encumbering conceptual developments with component-wise expressions, and from [11]. Interestingly, no matter how early in the history of elasticity the consequences of concentrated loads were studied, some of those, namely, the occurrence of concentrated contact interactions between adjacent body parte, went overlooked until recently [12–16].

2.1 Displacement, Strain, Compatibility

The problems in linear elasticity we are interested in are formulated over an unbounded region R of an Euclidean space $\mathcal{E}^N$ of dimension $N = 2$ or 3, R being either a half-space or the whole of $\mathcal{E}^N$; as a rule, in the following we take $N = 3$. Points x of R have a position vector

$$\boldsymbol{x} := x - o$$

with respect to a chosen point of $\mathcal{E}^N$, the *origin* o; the components of $\boldsymbol{x}$ in an orthonormal Cartesian basis $\boldsymbol{e}_i$ $(i = 1, 2, 3)$ are the Cartesian coordinates x_i:

$$\boldsymbol{x} = x_i \boldsymbol{e}_i .$$

P. Podio-Guidugli and A. Favata, *Elasticity for Geotechnicians*, Solid Mechanics and Its Applications 204, DOI: 10.1007/978-3-319-01258-2_2,

In this formula, we used *Einstein's convention*, consisting in leaving tacit the summation operation over the index range whenever in a monomial term an index is repeated twice: here, for example, this convention allows us to avoid the use of the more cumbersome notation

$$\boldsymbol{x} = \sum_{i=1}^{3} x_i \boldsymbol{e}_i .$$

Here and henceforth in this book we drop the qualifier 'orthogonal' for the only type of Cartesian coordinates we use.

In a deformation, a typical point $x \in R$ is displaced to a position

$$y = x + \boldsymbol{u}(x);$$

here, $\boldsymbol{u}$ is the vector field that describes the *displacement* from x to $y \in \mathcal{E}^N$. The *displacement gradient* is the tensor field whose value at x is by definition the outcome of taking the following limit:

$$\lim_{\varepsilon \to 0} \varepsilon^{-1}(\boldsymbol{u}(x + \varepsilon \boldsymbol{h}) - \boldsymbol{u}(x)) =: (\nabla \boldsymbol{u}(x))\boldsymbol{h}, \quad \forall \boldsymbol{h} \in \mathcal{V}, \tag{2.1}$$

where $\mathcal{V}$ is the N-dimensional vector space associated with $\mathcal{E}^N$. If $\boldsymbol{h}$ is a unit vector (that is, if $|\boldsymbol{h}| = 1$), the left side of the last relation defines the *directional derivative* of $\boldsymbol{u}$ in the direction $\boldsymbol{h}$:

$$\partial_{\boldsymbol{h}} \boldsymbol{u} := (\nabla \boldsymbol{u})\boldsymbol{h}.$$

On representing vector $\boldsymbol{u}$ in the chosen basis:

$$\boldsymbol{u} = u_i \boldsymbol{e}_i ,$$

an application of definition (2.1) yields the cartesian components of $\nabla \boldsymbol{u}$:

$$(\nabla \boldsymbol{u})_{ij} = u_{i,j},$$

where ,'$_j$' denotes differentiation with respect to coordinate x_j:

$$u_{i,j} = \frac{\partial u_i}{\partial x_j}.$$

Just as every other second-order tensor, $\nabla \boldsymbol{u}$ can be uniquely decomposed into the sum of its *symmetric part* $\boldsymbol{E}$ and its *skew-symmetric part* $\boldsymbol{W}$:

$$\begin{aligned}\nabla \boldsymbol{u} &= \boldsymbol{E}(\boldsymbol{u}) + \boldsymbol{W}(\boldsymbol{u}),\\ \boldsymbol{E}(\boldsymbol{u}) &:= \tfrac{1}{2}\big(\nabla \boldsymbol{u} + \nabla \boldsymbol{u}^T\big),\\ \boldsymbol{W}(\boldsymbol{u}) &:= \tfrac{1}{2}\big(\nabla \boldsymbol{u} - \nabla \boldsymbol{u}^T\big).\end{aligned} \tag{2.2}$$

The *strain tensor* $\boldsymbol{E}$ is a linear measure of the strain field associated with a given displacement field: its equal-index components ($E_{11} = u_{1,1}$ etc.) measure dilatation of fibers aligned with the Cartesian axes; the other components ($E_{12} = 1/2(u_{1,2} + u_{2,1})$ etc.) measure changes in the angle between fibers aligned along different axes; more generally, if $\boldsymbol{a}$ are $\boldsymbol{b}$ two mutually orthogonal unit vectors, $\boldsymbol{E}\boldsymbol{a}\cdot\boldsymbol{a}$ measures the dilatation of a fiber aligned with $\boldsymbol{a}$, and $\boldsymbol{E}\boldsymbol{a}\cdot\boldsymbol{b}(=\boldsymbol{E}\boldsymbol{b}\cdot\boldsymbol{a})$ measures the change in angle between fibers in the directions $\boldsymbol{a}$ and $\boldsymbol{b}$.[1]

The *rotation tensor* $\boldsymbol{W}$ furnishes a linear measure of the vorticity field associated with a given displacement field. The role of $\boldsymbol{W}$ is made clearer if the operation of taking the *curl* of $\boldsymbol{u}$ is introduced: this operation defines a vector field, denoted by $\operatorname{curl}\boldsymbol{u}$, such that

$$\boldsymbol{W}(\boldsymbol{u})\boldsymbol{a} =: \frac{1}{2}\operatorname{curl}\boldsymbol{u} \times \boldsymbol{a}, \quad \forall \boldsymbol{a} \in \mathcal{V}. \tag{2.3}$$

It follows from this definition that the Cartesian components of $\operatorname{curl}\boldsymbol{u}$ are:

$$(\operatorname{curl}\boldsymbol{u})_i = e_{ijk}u_{k,j}, \tag{2.4}$$

where e_{ijk} is *Ricci's symbol*.[2] We set:

---

[1] For more information about the role of $\boldsymbol{E}$ and, more generally, about the local analysis, both exact and approximate, of a deformation see [11], Chap. I

[2] In terms of the vectors composing the orthonormal Cartesian basis we chose, *Kronecker's symbol* δ_{ij} is given by

$$\delta_{ij} := \boldsymbol{e}_i \cdot \boldsymbol{e}_j,$$

whence

$$\delta_{ij} = \begin{cases} 1 \text{ if } & i = j \\ 0 \text{ if } & i \neq j \end{cases};$$

moreover, relation

$$e_{ijk} := \boldsymbol{e}_i \times \boldsymbol{e}_j \cdot \boldsymbol{e}_k$$

defines *Ricci's symbol*, so that

$$e_{ijk} = \begin{cases} +1 & \text{if all indices i, j, k are different and, in addition,}\\ & \text{their sequence is an even-class permutation of 1, 2, 3;}\\ 0 & \text{if at least two of the indices i, j, k are equal;}\\ -1 & \text{if all indices i, j, k are different and, in addition,}\\ & \text{their sequence is an odd-class permutation of 1, 2, 3.}\end{cases}$$

Ricci's and Kronecker's symbols are linked by the following relation:

$$e_{ijk}e_{lmk} = \delta_{il}\delta_{jm} - \delta_{im}\delta_{jl}. \tag{2.5}$$

By repeated saturation of pairs of free indices, two easy and often useful consequences of (2.5) are obtained:

$$\boldsymbol{w}(\boldsymbol{u}) := \frac{1}{2}\,\mathrm{curl}\,\boldsymbol{u};$$

this definition identifies $\boldsymbol{w}$ as the vector associated with $\boldsymbol{W}$ by the well-known one-to-one correspondence between $\mathcal{V}$ and Skw, the collection of all skew-symmetric second-order tensors, namely,

$$\mathcal{V} \ni \boldsymbol{v} \leftrightarrow \boldsymbol{V} \in \mathrm{Skw} \quad \Leftrightarrow \quad \boldsymbol{V}\boldsymbol{a} = \boldsymbol{v} \times \boldsymbol{a}, \quad \forall \boldsymbol{a} \in \mathcal{V}. \tag{2.6}$$

It is not difficult to show that

$$V_{ik} = e_{ijk} v_j, \quad v_i = \frac{1}{2} e_{ijk} V_{kj}.$$

In view of definition (2.1), we write:

$$\begin{aligned}\boldsymbol{u}(x) &= \boldsymbol{u}(x_0) + \boldsymbol{E}(x_0)(x - x_0) + \boldsymbol{W}(x_0)(x - x_0) + \boldsymbol{O}^2(|x - x_0|) \\ &= \boldsymbol{u}(x_0) + \boldsymbol{w}(x_0) \times (x - x_0) + \boldsymbol{E}(x_0)(x - x_0) + \boldsymbol{O}^2(|x - x_0|),\end{aligned}$$

where $\boldsymbol{E}(x_0) = \boldsymbol{E}(\boldsymbol{u}(x_0))$ etc. The last equality makes clear what is meant by *local linear approximation* of a given displacement field $\boldsymbol{u}$, that is, by the approximation of $\boldsymbol{u}$ to within terms of order $\boldsymbol{O}^2(|x - x_0|)$ in a neighbourhood of an arbitrarily chosen interior point x_0 of R): it consists of the sum of a *rigid displacement*

$$\boldsymbol{u}_{rig}(x) := \boldsymbol{u}(x_0) + \boldsymbol{w}(x_0) \times (x - x_0),$$

made up of a *translation* $\boldsymbol{u}(x_0)$ and of a *rotation* about x_0 of vector $\boldsymbol{w}(x_0)$, and of a nonrigid displacement

$$\boldsymbol{u}_{def}(x) = \boldsymbol{E}(x_0)(x - x_0),$$

the only part of $\boldsymbol{u}$ inducing what in everyday language is called a 'small deformation'. In fact, $\boldsymbol{E}$ is often called the *infinitesimal strain tensor*, the modifier 'strain' being an alternative to 'deformation' and the modifier 'infinitesimal' being used to distinguish $\boldsymbol{E} = \mathrm{sym}(\nabla \boldsymbol{u})$ from other local measures of deformation that, being exact, depend nonlinearly on $\nabla \boldsymbol{u}$.

We introduce here some more notions to be used in what follows.

Lin is the space of all second-order tensors, regarded as linear transformations of $\mathcal{V}$ into itself; Sym and Skw are two complementary subspaces of Lin, respectively, the

---

(i) formal multiplication of both sides by δ_{jm} yields:

$$e_{ijk} e_{ljk} = 2\,\delta_{il}\,;$$

(ii) one more saturation gives:

$$e_{ijk} e_{ijk} = 6\,.$$

subspace of symmetric ($\boldsymbol{A} = \boldsymbol{A}^T$) and skew-symmetric ($\boldsymbol{A} = -\boldsymbol{A}^T$) tensors. When $\dim(\mathcal{V}) = 3$, $\dim(\text{Lin}) = 9$, $\dim(\text{Sym}) = 6$ and $\dim(\text{Skw}) = 3$; when $\dim(\mathcal{V}) = 2$, $\dim(\text{Lin}_{(2)}) = 4$, $\dim(\text{Sym}_{(2)}) = 3$ and $\dim(\text{Skw}_{(2)}) = 1$.

Remark 2.1 With the use of (2.6), it can be shown that the vector associated with the skew-symmetric tensor $(\boldsymbol{a}\otimes\boldsymbol{b} - \boldsymbol{b}\otimes\boldsymbol{a})$ is $\boldsymbol{b}\times\boldsymbol{a}$.[3] Every skew-symmetric tensor can be represented as the linear combination of the following tensors:

$$\begin{aligned} \boldsymbol{W}_1 &= -\boldsymbol{e}_2 \otimes \boldsymbol{e}_3 + \boldsymbol{e}_3 \otimes \boldsymbol{e}_2, \\ \boldsymbol{W}_2 &= -\boldsymbol{e}_3 \otimes \boldsymbol{e}_1 + \boldsymbol{e}_1 \otimes \boldsymbol{e}_3, \\ \boldsymbol{W}_3 &= -\boldsymbol{e}_1 \otimes \boldsymbol{e}_2 + \boldsymbol{e}_2 \otimes \boldsymbol{e}_1, \end{aligned} \tag{2.7}$$

where

$$\boldsymbol{W}_i \leftrightarrow \boldsymbol{e}_i .$$

Remark 2.2 The *divergence* of a vector field $\boldsymbol{u}$ is the scalar field

$$\operatorname{div}\boldsymbol{u} := \operatorname{tr}(\nabla\boldsymbol{u});$$

it follows from this definition that

$$\operatorname{div}\boldsymbol{u} = \operatorname{tr}\boldsymbol{E}(\boldsymbol{u}) = E_{ii} = u_{i,i} .$$

Note that

$$\operatorname{div}\operatorname{curl}\boldsymbol{u} = 0,$$

and that, for φ a scalar field,

$$\operatorname{curl}\nabla\varphi = \boldsymbol{0} \quad \text{and} \quad \operatorname{div}\nabla\varphi = \Delta\varphi. \tag{2.8}$$

These two identities help to interpret a classical result in vector calculus, *Helmholtz's Decomposition Theorem*:

> given any sufficiently smooth field $\boldsymbol{u}$ over a bounded regular region R, there are a scalar field φ and a divergenceless vector field $\boldsymbol{w}$ over R such that
>
> $$\boldsymbol{u} = \nabla\varphi + \operatorname{curl}\boldsymbol{w};$$

if $\boldsymbol{u} \in C(\bar{R}) \cap C^M(R)$, $M \geq 1$, then both φ and $\boldsymbol{w}$ are of class $C^M(R)$.

Note that a straightforward application of (2.8) yields:

$$\operatorname{curl}\boldsymbol{u} = \operatorname{curl}\operatorname{curl}\boldsymbol{w} \quad \text{and} \quad \operatorname{div}\boldsymbol{u} = \Delta\varphi.$$

---

[3] Recall that the symbol $\otimes$ signifies dyadic product, a notion introduced in the first footnote of Sect. 1.3; the second-order tensor $\boldsymbol{a}\otimes\boldsymbol{b}$ is defined by specifying its linear action on vectors.

2.1.1 Compatibility

With each displacement field $\boldsymbol{u}$ of class $C^1(R)$ we can always associate a continuous deformation field $\boldsymbol{E}$ such that

$$2\boldsymbol{E} = \nabla\boldsymbol{u} + \nabla\boldsymbol{u}^T; \tag{2.9}$$

in components,

$$2E_{ij} = u_{i,j} + u_{j,i}. \tag{2.10}$$

This relation can also be regarded as the tensorial equation ruling the problem of finding a displacement field $\boldsymbol{u}$ associated with a given strain field $\boldsymbol{E}$. This problem is *overdetermined*, because the three unknown fields u_i are restricted by the six scalar equations (2.10). Not that problems of this type have necessarily no solution. However, for them the *well-posedness* issue (**a**. Are there solutions? **b**. If answer to **a** is yes, how many are they? **c**. Do solutions depend continuously on data?) can be discussed only after having checked that the assigned data satisfy certain a priori solvability conditions called *compatibility conditions*. We now deduce such conditions for the case of our current interest.

To begin with, we have to put together a curl notion for tensor-valued fields. We do so by exploiting the definition given in (2.3) for vector-valued fields:

$$(\mathrm{curl}\,\boldsymbol{A})\boldsymbol{a} := \mathrm{curl}\,(\boldsymbol{A}^T\boldsymbol{a}), \quad \forall \boldsymbol{a} \in \mathcal{V};$$

in components,

$$(\mathrm{curl}\,\boldsymbol{A})_{ij} = e_{ipq}A_{jq,p}.$$

If we now apply formally the operator curl on both sides of (2.9), we find[4]:

$$2\,\mathrm{curl}\,\boldsymbol{E} = \mathrm{curl}\,(\nabla\boldsymbol{u}) + \mathrm{curl}\,(\nabla\boldsymbol{u}^T) = \mathrm{curl}\,(\nabla\boldsymbol{u}^T) = 2\nabla\boldsymbol{w}. \tag{2.11}$$

Taking the curl of (2.11), we arrive at the sought-for compatibility condition:

$$\mathrm{curl}\,\mathrm{curl}\,\boldsymbol{E} = \boldsymbol{0}; \tag{2.12}$$

in components,

[4] That curl $(\nabla\boldsymbol{u}) = \boldsymbol{0}$ follows from the definitions of (the two involved operators and) Ricci symbol:

$$(\mathrm{curl}\,(\nabla\boldsymbol{u}))_{ij} = e_{ipq}(\nabla\boldsymbol{u})_{jq,p} = e_{ipq}(u_{j,q})_{,p} = e_{ipq}u_{j,qp} = 0.$$

Furthermore, in view of (2.4),

$$(\mathrm{curl}\,(\nabla\boldsymbol{u}^T))_{ij} = e_{ipq}(u_{q,j})_{,p} = e_{ipq}u_{q,jp} = (e_{ipq}u_{q,p})_{,j} = 2\,w_{i,j}.$$

$$e_{ijk}e_{lmn}E_{jm,kn} = 0. \tag{2.13}$$

If region R is simply connected, for each given symmetric-valued field $\boldsymbol{E}$ of class $C^K(R)$, $K \geq 2$ there is a class $C^{K+1}(R)$ displacement field $\boldsymbol{u}$, which satisfies (2.9).[5] The field $\boldsymbol{u}$ can be constructed by means of *Cesàro's formula*:

$$u_i(x) = \int_{x_0}^{x} U_{ij}(y,x)dy_j, \quad U_{ij}(y,x) := E_{ij}(y) + (x_k - y_k)(E_{ij},_k(y) - E_{kj},_i(y)), \tag{2.14}$$

where the integral does not depend on the path that has been chosen in R to connect a given point x_0 with the typical point x. Needless to say, this formula determines $\boldsymbol{u}$ to within an arbitrary rigid displacement.

Remark 2.3 The representation (1.8) for the displacement field in an elastic beam subject solely to axial loads can be regarded as a minimal version of this general formula: for $\boldsymbol{e}$ a unit vector parallel to the axis, the strain field is

$$\boldsymbol{E}(z) = w'(z)\boldsymbol{e} \otimes \boldsymbol{e},$$

whence, by (2.14), $\boldsymbol{U}(z, \zeta) \equiv \boldsymbol{E}(\zeta)$ and

$$\boldsymbol{u}(z) = \int_{z_0}^{z} (\boldsymbol{U}(\zeta)\boldsymbol{e})\, d\zeta = \left(\int_{z_0}^{z} w'(\zeta)d\zeta\right)\boldsymbol{e}.$$

2.1.2 Plane Displacement Fields

A displacement field $\boldsymbol{u}$ is called *plane* whenever there is a Cartesian reference with respect to which $\boldsymbol{u}$ admits the representation:

$$u_\alpha = u_\alpha(x_1, x_2),\ \alpha = 1, 2, \quad u_3 \equiv 0, \tag{2.15}$$

at any point $x \in R$.[6] The corresponding strain state is:

$$E_{\alpha\beta} = \frac{1}{2}(u_{\alpha},_{\beta} + u_{\beta},_{\alpha}), \quad E_{3i} \equiv 0$$

(compare with (2.10)).

---

[5] For a proof of this result, which is due to the great Italian elasticist Eugenio Beltrami (1835–1900), who established it in 1889, see [6], Sect. 14, where various other results included in this section are also proved.

[6] When Greek indices are used, it is understood that they take the values 1 and 2; the range of Latin indices is the set $\{1, 2, 3\}$.

Remark 2.4 For an example of plane displacement field, consider the rigid displacement:

$$\boldsymbol{v} = \boldsymbol{t} + \alpha \boldsymbol{W}_3 \boldsymbol{x} = \boldsymbol{t} + \alpha\, \boldsymbol{e}_3 \times \boldsymbol{x}, \qquad \boldsymbol{t} = t_\alpha \boldsymbol{e}_\alpha, \quad \boldsymbol{x} = x_\alpha \boldsymbol{e}_\alpha, \tag{2.16}$$

consisting of a rotation of α radians about an axis of unit vector $\boldsymbol{e}_3$ and of a translation $\boldsymbol{t}$ in the plane perpendicular to $\boldsymbol{e}_3$. It is easy to see that each field $\boldsymbol{v}$ of type (2.16) solves the following differential system:

$$v_{1,1} = 0, \quad v_{2,2} = 0, \quad v_{1,2} + v_{2,1} = 0; \tag{2.17}$$

as a matter of fact, in components relations (2.16) read:

$$v_1 = t_1 - \alpha\, x_2, \quad v_2 = t_2 + \alpha\, x_1. \tag{2.18}$$

If a rigid plane field whatsoever is added to any plane deformation field, the relative strain state stays the same.

2.1.3 Plane Strain Fields

A strain field $\boldsymbol{E}$ is called *plane* whenever its component representation in a suitable Cartesian reference is:

$$E_{\alpha\beta} = E_{\alpha\beta}(x_1, x_2), \quad E_{3i} \equiv 0. \tag{2.19}$$

For such a field, the tensorial compatibility condition (2.12) shrinks to one scalar relation:

$$2\, E_{12,12} = E_{11,22} + E_{22,11}; \tag{2.20}$$

interestingly, of the six conditions (2.13) this is the one obtained when both free indices are taken equal to 3.

Remark 2.5 For plane strain fields, Cesàro's formula gives:

$$u_\alpha(x) = \int_{x_0}^{x} U_{\alpha\beta}(y, x)\, dy_\beta,$$
$$U_{\alpha\beta}(y, x) = E_{\alpha\beta}(y) + (x_\gamma - y_\gamma)(E_{\alpha\beta,\gamma}(y) - E_{\gamma\beta,\alpha}(y)).$$

The strain field associated with a plane displacement field is plane. We proceed to give a direct proof of the converse statement. To begin with, a displacement field $\boldsymbol{u}$ satisfying the last three relations (2.19) must be such that

$$u_{3,\alpha} + u_{\alpha,3} = 0 \tag{2.21}$$

and that

$$u_{3,3} = 0,$$

that is, such that u_3 be independent of x_3:

$$u_3 = u_3(x_1, x_2). \tag{2.22}$$

Relations (2.21) and (2.22) imply that

$$u_{\alpha,33} = 0,$$

or rather, equivalently, that

$$u_\alpha = \widehat{u}_\alpha(x_1, x_2) + x_3\,\widehat{v}_\alpha(x_1, x_2). \tag{2.23}$$

On combining this preliminary representation for u_α with what the first three relations (2.19) require (namely, that each of the components $E_{\alpha\beta}$ of $\boldsymbol{E}$ be indipendent of x_3), we infer that the vector field $\boldsymbol{v}$ must obey the differential relations (2.17), and hence that it must have the form (2.18); we then set:

$$v_1 = a_1 - b\,x_2, \quad v_2 = a_2 + b\,x_1. \tag{2.24}$$

At this point, we insert representations (2.22), (2.23) and (2.24) into relations (2.21), so as to obtain:

$$u_{3,1} + a_1 - b\,x_2 = 0, \quad u_{3,2} + a_2 + b\,x_1 = 0, \tag{2.25}$$

whence by differentiation we deduce that

$$u_{3,12} - b = 0, \quad u_{3,21} + b = 0,$$

that is,

$$b = 0, \quad u_{3,12} = 0.$$

With the use of the first result, we achieve a preliminary representation, more precise than (2.23), for functions u_α:

$$u_\alpha = \widehat{u}_\alpha(x_1, x_2) + a_\alpha x_3;$$

The definitive form we choose for such representation is:

$$\begin{aligned} u_1 &= \breve{u}_1(x_1, x_2) + t_1 - a_3\,x_2 + a_1 x_3, \\ u_2 &= \breve{u}_2(x_1, x_2) + t_2 + a_3\,x_1 + a_2 x_3, \end{aligned} \tag{2.26}$$

where $\breve{\boldsymbol{u}}$, the part of $\widehat{\boldsymbol{u}}$ responsible for shape and/or volume changes, is distinguished from the rigid part, the latter being of type (2.18). It is easy to check that the plane field $\boldsymbol{E}(\breve{\boldsymbol{u}})$ satisfies (2.20).

Now, given that $b = 0$, relations (2.25) have the following consequences:

$$(u_{3,1} + a_1 = 0 \Rightarrow) \;\; -a_1x_1 + c_1(x_2) = u_3 = -a_2x_2 + c_2(x_1) \;\; (\Leftarrow u_{3,2} + a_2 = 0).$$

This double expression for u_3 holds true for arbitrary values of the independent variables x_1, x_2 provided

$$a_1x_1 + c_2(x_1) = a_2x_2 + c_1(x_2) = t_3,$$

with t_3 an arbitrary constant; hence,

$$u_3(x_1, x_2) = t_3 - (a_1x_1 + a_2x_2).$$

This expression is found compatible with (2.21) and (2.26) if $a_1 = a_2 = 0$.

In conclusion, given a plane strain field as in (2.19), the corresponding displacement field consists of a plane field $\breve{\boldsymbol{u}}$ such that

$$\breve{u}_{\alpha,\beta} + \breve{u}_{\beta,\alpha} = 2E_{\alpha\beta}$$

and of a rigid displacement field featuring an arbitrary translation and an arbitrary small rotation about the third axis:

$$\boldsymbol{r} = \boldsymbol{t} + \boldsymbol{A}\boldsymbol{x}, \quad \boldsymbol{A} = -\boldsymbol{A}^T, \quad \boldsymbol{x} = x_\alpha \boldsymbol{e}_\alpha, \tag{2.27}$$

where, on recalling $(2.7)_3$, $\boldsymbol{A} = -a_3(\boldsymbol{e}_1 \otimes \boldsymbol{e}_2 - \boldsymbol{e}_2 \otimes \boldsymbol{e}_1) = a_3\boldsymbol{W}$.

2.2 Forces, Stress, Equilibrium

In continuum mechanics, a body is generally thought of as subject to *distance* and *contact actions* on the part of its environment. No matter in what placement in physical space a body is observed, both types of actions are customarily modeled as *diffuse*: those at a distance as forces per unit volume, just as is done in the familiar case of gravity; contact actions as forces per unit surface, on the basis of examples like the pressure exerted by a fluid on a body immersed into it (the wind on a sail) or containing it (the water on a glass).

In most cases, distance actions between disjoint parts of the same body are neglected, as are the distance actions of a part on itself (e.g., self-gravitation). Distance actions at a typical interior body point x are specified by the value taken at that point by an assigned vector field $\widehat{\boldsymbol{d}}$; they are customarily split into *inertial* and *noninertial* parts:

$$\widehat{\boldsymbol{d}}(x) = \widehat{\boldsymbol{d}}^{in}(x) + \widehat{\boldsymbol{d}}^{ni}(x), \quad \widehat{\boldsymbol{d}}^{in}(x) := -\rho(x)\ddot{\boldsymbol{x}},$$

where $\rho(x)$ is the current *mass density*, and $\ddot{\boldsymbol{x}}$ the *acceleration*, at x. In this book, we shall never consider bodies in motion, and hence there will be no need to worry about inertial forces.

In all cases, in addition to contact interactions of a body with its environment, adjacent body parts are presumed to have diffuse *contact interactions*, which are thought of as accounting for the short-range forces between neighboring particles envisaged by discrete mechanics. Mathematically, such contact interactions are described by a vector field $\widehat{\boldsymbol{c}}(\cdot,\cdot)$ defined over the Cartesian product of the body's closure times the sphere of unit vectors: when evaluated at a point x of a common boundary surface oriented by the unit normal $\widehat{\boldsymbol{n}}(x)$, such so-called *stress-vector* field is interpreted as delivering the force $\widehat{\boldsymbol{c}}(x,\widehat{\boldsymbol{n}}(x))$ per unit area exerted either by the environment over the body or by the part lying on the positive side of the boundary surface over the adjacent part.[7]

Concentrated external actions, under form of forces applied at interior or boundary points, have also been considered; their mechanical effects are of central interest in this book. As we shall see, when applied at a boundary point—as is the case with the Flamant Problem we study in Chap. 4—they were regarded as limits of distributions of contact actions localized in a surface neighborhood of that point, which was made to shrink to null; similarly, when applied to an interior point, as in the case of Kelvin Problem to be studied in Chap. 5, they were regarded as limits of distributions of distance actions localized in a volume neighborhood of that point. Surprisingly enough, the occurrence of *concentrated contact interactions between adjacent body parts* went noticed until recently, when Flamant's and other problems of the same type were re-examined [12] (see also [13]).[8]

2.2.1 Cauchy's Notion of Stress

A body acted upon by a force system $(\boldsymbol{d}, \boldsymbol{c})$ is said to occupy an *equilibrium placement* $\mathcal{B}$ when it so happens that

---

[7] It appears that the concept of diffused contact interactions between internal adjacent body parts begun to condensate in Cauchy's mind on the basis of a similarity with standard examples of diffused contact loads exerted on a body by an environment of a different nature, such as the hydrostatic pressure of a fluid on an immersed solid [3]. Cauchy's model of internal contact interactions has been applied without changes to contact interactions of a body with its exterior, with the stress-vector mapping accounting for both. An implicit drawback of this practice is that no difference is made between geometrical surfaces obtained by ideal cuttings and fabricated surfaces obtained by actual cuttings [4]; moreover, the issue of boundary compatibility of a (body,environment) pair is completely overlooked [1, 2].

[8] The construction of an interaction theory general enough to allow for concentrated contact interactions between adjacent body parts has been undertaken by Schuricht [15, 16]; among the intriguing features of such a theory is the rethinking it involves of the body-part notion. In [14], examples are given of interactions in cuspidate bodies that concentrate at the cusp point, regarded as a body part.

$$\int_{\mathcal{P}} \boldsymbol{d}\cdot\boldsymbol{r}+\int_{\partial\mathcal{P}} \boldsymbol{c}\cdot\boldsymbol{r}=\mathbf{0}, \tag{2.28}$$

for all parts $\mathcal{P}$ of $\mathcal{B}$ and for all rigid fields $\boldsymbol{r}$ as in (2.27) (here, as anticipated, $\boldsymbol{d}$ stands for the noninertial distance force). By virtue of *Cauchy's Stress Theorem* (see, e.g., [7], Sect. 14), it follows from (2.28), when written for an arbitrary translation $\boldsymbol{t}$, that the stress-vector mapping can be represented as follows:

$$\widehat{\boldsymbol{c}}(x,\boldsymbol{n})=\widehat{\mathbf{S}}(x)\boldsymbol{n}. \tag{2.29}$$

in terms of a *stress-tensor* field $\widehat{\mathbf{S}}$ defined over the closure of $\mathcal{B}$: the affine action of $\widehat{\mathbf{S}}(x)$ over the sphere of unit vectors yields the stress vector on the triple infinity of oriented planes through x. Conversely, given the stress-vector mapping $\widehat{\boldsymbol{c}}(x,\cdot)$ at a typical body point x and three mutually orthogonal unit vectors $\boldsymbol{n}_i$, the construct

$$\widehat{\mathbf{S}}(x)=\sum_{i=1}^{3}\widehat{\boldsymbol{c}}(x,\boldsymbol{n}_i)\otimes\boldsymbol{n}_i \tag{2.30}$$

defines the value at x of the stress-tensor field. Thus—and this is the main thrust of Cauchy's result—*the information carried by the stress-vector and stress-tensor mappings* $\widehat{\boldsymbol{c}}$ *and* $\widehat{\mathbf{S}}$ textitare essentially equivalent.

It follows from (2.28) and (2.29) that

$$\int_{\mathcal{P}} \boldsymbol{d}+\int_{\partial\mathcal{P}} \mathbf{S}\boldsymbol{n}=\boldsymbol{0},\quad \forall\mathcal{P}\subset\mathcal{B},$$

whence, granted regularity,

$$\operatorname{div}\mathbf{S}+\boldsymbol{d}=\boldsymbol{0}\ \text{ in } \mathcal{B}. \tag{2.31}$$

Moreover, it follows from (2.31) and (2.28), when written for an arbitrary rotation $\boldsymbol{A}$, that the stress field is symmetric-valued:

$$\mathbf{S}=\mathbf{S}^T.$$

2.2.2 Free-Body Diagrams, Diffuse and Concentrated Forces

A feature of the equilibrium statement (2.28)—namely, that whatever part of an equilibrated body must be in equilibrium as well—would be hardly contended by anybody. The widespread and fruitful use of *free-body diagrams* in mechanics is based on this assumption, and on the accompanying presumption that a body part, when ideally isolated from the rest by a so-called *Euler cut*, would be in equilibrium if it were acted upon by external forces reproducing faithfully the forces, both external and internal, it directly experiences in reality. Usually, the subbodies whose equilibrium is characterized in this manner are imagined to have an everywhere smooth boundary. Not always so in this book, where consideration of sharp-cornered parts is at times necessary to exhibit the concentrations of contact forces that at times may occur (see e.g. Fig. 4.8).

Concentrated forces, regarded as convenient idealizations of diffused loads applied to a small part of a body's boundary, are of common use in engineering mechanics. To quote from a popular textbook, "the free-body diagram is the most important single step in the solution of problems in mechanics" ([10], p. 104); "modeling the action of forces" "exerted *on* the body to be *isolated*, *by* the body to be *removed*" (*ibid.*, p. 105; italics as in the original text) is a mandatory, preliminary step; and those forces, especially but not exclusively in statics, are for most practical purposes modeled as concentrated.

Strictly speaking, the equivalence in information content of (2.30) and (2.29) holds true for *diffused* contact force and *regular* stress fields. In the next chapters, we display and discuss situations when *concentrated* contact forces and *singular* stress fields are in order. Precisely, first by inspection of a problem of pure statics, which is the two-dimensional counterpart of the Flamant problem, then by inspection of the three-dimensional problem Flamant solved, as well as those solved by Boussinesq, Cerruti and Kelvin, we demonstrate *per exempla* that *partwise equilibrium of a simple continuous body may require that adjacent body parts exchange concentrated contact forces*.

We have seen that diffused contact loads are germane to contact interactions between adjacent body parts, so much so that they are customarily described by one and the same vector-valued mapping. Concentrated loads, applied at interior and boundary points, have been often considered in continuum mechanics, and carefully modeled mathematically (for the class of linearly elastic bodies, see [6], Sect. 52). We see no reason why the germane notion of concentrated contact interactions should not be introduced. They are *not* ubiquitous; in fact, they are a rather rare necessity. Let us revert for a moment to engineering mechanics for guidance. A judicious practice there is to make sure that the free-body diagram features *all* possible forces applied to the isolated body; at times, we find out that balance and/or symmetry conditions require that some of those forces be null. Likewise, in continuum mechanics, we should contemplate concentrated contact interactions by default, because there are cases, no matter how few, when they turn out to be crucial to guarantee partwise equilibrium.

If concentrated contact interactions are considered, an interesting problem to tackle is the conjectural equivalence in information of contact forces, regular and singular, and the accompanying, somewhere singular, stress field. Luckily, *concentrated forces occur 'naturally' in weak formulations of force-balance laws*, be they idealizations of applied loads or of contact interactions. In fact, in such formulations, concentrated loads are as 'natural' as edges and vertices in the domain where a boundary value problem is formulated. There is no need today to justify consideration of concentrated forces, as was done over a century ago, by thinking of them as limits of smooth distributions of volume or surface forces, just as there is no need to round off a domain's corners. In addition, weak formulations relieve us from dealing with a delicate issue arising when sequences of approximating problems are employed, namely, to investigate under what hypotheses an associated sequence of smooth solutions has a unique limit.

2.3 The Stress Response to Strain

In classical elasticity, the stress response to strain is described by a linear mapping of the collection of all symmetric tensors into itself:

$$\mathrm{Sym} \ni \boldsymbol{E} \mapsto \boldsymbol{S} \in \mathrm{Sym}, \quad \boldsymbol{S} = \mathbb{C}\boldsymbol{E} \quad (S_{ij} = \mathbb{C}_{ijhk}E_{hk}), \tag{2.32}$$

where the *elasticity tensor* $\mathbb{C}$ has the following index-pair symmetries:

$$\mathbb{C}_{ijhk} = \mathbb{C}_{jihk} = \mathbb{C}_{ijkh}, \quad \mathbb{C}_{ijhk} = \mathbb{C}_{hkij}.$$

Collectively, these symmetries guarantee that

(i) all of the $3^4 = 81$ Cartesian components of $\mathbb{C}$ are expressible in terms of only 21 of them, in general mutually independent;
(ii) there is a quadratic scalar-valued function defined over Sym:

$$\mathrm{Sym} \ni \boldsymbol{E} \mapsto \sigma \in \mathbb{R}, \quad \sigma = \sigma(\boldsymbol{E}) = \frac{1}{2}\boldsymbol{E} \cdot \mathbb{C}\boldsymbol{E} = \frac{1}{2}\mathbb{C}_{ijhk}E_{ij}E_{hk}, \tag{2.33}$$

referred to as the *strain energy* per unit referential volume, such that

$$\partial_{\boldsymbol{E}}\sigma(\boldsymbol{E}) = \mathbb{C}\boldsymbol{E}.$$

It follows from (2.32) and, respectively, (2.33) that, $\boldsymbol{S} = \boldsymbol{0}$ and $\sigma = 0$ for $\boldsymbol{E} = \boldsymbol{0}$. It is when both the stress and the strain energy are null at a point—that is, when the material is in a *natural state* at that point—that classical elasticity studies the local response of a linearly elastic material to the various causes of deformation. For reasons of physical plausibility, the strain energy is assumed to be *positive definite*, i.e., such that

$$\sigma(\boldsymbol{E}) \geq 0, \quad \sigma(\boldsymbol{E}) = 0 \Leftrightarrow \boldsymbol{E} = \boldsymbol{0}. \tag{2.34}$$

This assumption is more than sufficient to guarantee that the constitutive mapping (2.32) be invertible:

$$\boldsymbol{E} = \mathbb{C}^{-1}\boldsymbol{S}. \tag{2.35}$$

2.3.1 Isotropic Materials

When a material's response is "the same in whatever direction", that material is said *isotropic*. The elasticity tensor of an isotropic linearly elastic material is completely determined by two parameters only, the so-called *Lamé's moduli* λ and μ; the stress-strain law has the following form:

$$\boldsymbol{S} = 2\mu\boldsymbol{E} + \lambda(\operatorname{tr}\boldsymbol{E})\boldsymbol{I}, \quad S_{ij} = 2\mu E_{ij} + \lambda(E_{hh})\delta_{ij} \tag{2.36}$$

(here $\boldsymbol{I}$ denotes the identity tensor), while the strain energy reads:

$$\sigma(\boldsymbol{E}) = \mu\,|\boldsymbol{E}|^2 + \frac{1}{2}\lambda(\operatorname{tr}\boldsymbol{E})^2 = \mu\,E_{ij}E_{ij} + \frac{1}{2}\lambda(E_{hh})^2;$$

for (2.34) to hold, it is necessary and sufficient that

$$\mu > 0, \quad 3\lambda + 2\mu > 0. \tag{2.37}$$

It is not difficult to determine the form taken by the inverse constitutive equation (2.35). Firstly, on taking the trace of (2.36), one obtains that

$$\operatorname{tr}\boldsymbol{S} = (3\lambda + 2\mu)\operatorname{tr}\boldsymbol{E}; \tag{2.38}$$

next, in view also of (2.37), one arrives at:

$$\boldsymbol{E} = \frac{1}{2\mu}\Big(\boldsymbol{S} - \frac{\lambda}{3\lambda + 2\mu}(\operatorname{tr}\boldsymbol{S})\boldsymbol{I}\Big). \tag{2.39}$$

Remark 2.6 For isotropic materials, the equilibrium equation (2.31) can be written in terms of displacement as Louis Navier (1785–1836) did first:

$$\mu\Delta\boldsymbol{u} + (\lambda + \mu)\nabla(\operatorname{div}\boldsymbol{u}) + \boldsymbol{d} = \boldsymbol{0}. \tag{2.40}$$

In this equation, three differential operators appear: laplacian and divergence of a vector field, and gradient of a scalar field. On recalling how these operators look like in Cartesian components[9]:

$$(\Delta\boldsymbol{v})_i = v_{i,jj}, \quad \operatorname{div}\boldsymbol{v} = v_{i,i}, \quad \text{and} \quad (\nabla\varphi)_i = \varphi_{,i},$$

[9] The laplacian of a vector field $\boldsymbol{v}$ is the vector field that obtains by taking the divergence of the gradient of $\boldsymbol{v}$:

$$\Delta\boldsymbol{v} = \operatorname{div}(\nabla\boldsymbol{v});$$

its Cartesian components have the form just shown because $(\nabla\boldsymbol{v})_{ij} = v_{i,j}$ and because, for $\boldsymbol{V}$ a second-order tensor field, $(\operatorname{div}\boldsymbol{V})_i = V_{ij,j}$.

the component version of Navier equation is easy to write:

$$\mu u_{i,jj} + (\lambda + \mu) u_{j,ji} + d_i = 0.$$

Remark 2.7 Let the distance forces be null. Then, on taking the divergence of Navier equation, one finds:

$$(\lambda + 2\mu)\Delta(\operatorname{div}\boldsymbol{u}) = 0,$$

whence, given that

$$\operatorname{div}\boldsymbol{u} = \operatorname{tr}\boldsymbol{E}(\boldsymbol{u})$$

and that it follows from (2.37) that

$$\lambda + 2\mu > 0,$$

one obtains

$$\Delta(\operatorname{tr}\boldsymbol{E}(\boldsymbol{u})) = 0.$$

But, if tr $\boldsymbol{E}(\boldsymbol{u})$ has to be a *harmonic* function (that is, a function whose laplacian is null), then tr $\boldsymbol{S}(\boldsymbol{u})$ must be harmonic as well, because of (2.38).[10] We shall deduce this condition again, in a different manner, in Sect. 2.4, where we study the compatibility issue in terms of stresses.

2.3.2 Mechanical Interpretation of the Elastic Moduli

The role of the elastic moduli is clarified when one imagines to perform some typical experiments, in each of which the one or the other modulus enters in a perspicuous manner. In the first two experiments we are going to consider, we record what stress accompanies a given strain according to the constitutive relation (2.36); in the third one, the stress is assigned, and the corresponding strain is computed with the use of (2.39).

(a) Simple shearing
For $\boldsymbol{a}, \boldsymbol{b}$ two orthogonal vectors,

$$\boldsymbol{E} = \tau(\boldsymbol{a}\otimes\boldsymbol{b} + \boldsymbol{b}\otimes\boldsymbol{a}) \quad \Rightarrow \quad \boldsymbol{S} = \tau\, 2\mu(\boldsymbol{a}\otimes\boldsymbol{b} + \boldsymbol{b}\otimes\boldsymbol{a});$$

therefore,

$$2\mu := \frac{\boldsymbol{b}\cdot\boldsymbol{S}\boldsymbol{a}}{\boldsymbol{b}\cdot\boldsymbol{E}\boldsymbol{a}},$$

[10] Here, $\boldsymbol{S}(\boldsymbol{u}) := 2\mu\boldsymbol{E}(\boldsymbol{u}) + \lambda(\operatorname{tr}\boldsymbol{E}(\boldsymbol{u}))\boldsymbol{I}$.

the *shear modulus*, measures the shear stress necessary to sustain a unit shearing strain.

(b) Uniform dilatation

$$\boldsymbol{E} = \tau \boldsymbol{I} \quad \Rightarrow \quad \boldsymbol{S} = \tau(3\lambda + 2\mu)\boldsymbol{I};$$

hence, the *dilatation modulus*:

$$3\lambda + 2\mu := \frac{\boldsymbol{S} \cdot \boldsymbol{I}}{\boldsymbol{E} \cdot \boldsymbol{I}} \tag{2.41}$$

is proportional to the *pressure* $1/3(\boldsymbol{S} \cdot \boldsymbol{I})$ accompanying the *volume change* $\boldsymbol{E} \cdot \boldsymbol{I}$.

(c) Uniaxial stress

Again, let $\boldsymbol{a}$ e $\boldsymbol{b}$ be two orthogonal vectors. Then,

$$\boldsymbol{S} = \tau \boldsymbol{a} \otimes \boldsymbol{a} \quad \Rightarrow \quad \boldsymbol{E} = \tau \frac{1}{2\mu}\left(\boldsymbol{a} \otimes \boldsymbol{a} - \frac{\lambda}{3\lambda + 2\mu}\boldsymbol{I}\right).$$

The *Young's modulus*

$$E := \frac{\boldsymbol{a} \cdot \boldsymbol{S}\boldsymbol{a}}{\boldsymbol{a} \cdot \boldsymbol{E}\boldsymbol{a}} = \frac{\mu(3\lambda + 2\mu)}{\lambda + \mu}$$

measures the axial stress necessary to cause a unit axial strain. The *Poisson's modulus* (also known as the *lateral-contraction modulus*)

$$\nu := -\frac{\boldsymbol{b} \cdot \boldsymbol{E}\boldsymbol{b}}{\boldsymbol{a} \cdot \boldsymbol{E}\boldsymbol{a}} = \frac{\lambda}{2(\lambda + \mu)}$$

measures the transverse-to-axial strain ratio in an experiment where an axial stress state is induced. The moduli E, ν and

$$G := \mu$$

are those currently used in the (geo)technical applications of linear and isotropic elasticity. We also note for later reference another expression for the dependence of volume changes on pressure:

$$\operatorname{tr}\boldsymbol{E} = \frac{1 - 2\nu}{E}\operatorname{tr}\boldsymbol{S} \tag{2.42}$$

(cf. (2.41)).

Remark 2.8 As Lamé's constitutive equation shows, two moduli characterize completely the response of an isotropic material. In fact, it is not difficult to see that the three technical moduli are linked by the consistency condition

$$E = 2(1+\nu)G. \tag{2.43}$$

Remark 2.9 The positivity inequalities (2.37) imply that

$$E,\, G > 0, \quad -1 < \nu < 1/2\,. \tag{2.44}$$

Therefore, linearly elastic and isotropic materials that contract transversely when axially extended (that is, materials for which $0 < \nu < 1/2$) have an E/G ratio strictly included between 2 and 3; and, for those whose $\nu \in (-1, -1/2)$, to have a Young's modulus smaller (even much smaller) that their shear modulus does not forbid the strain energy to be positive definite.

Remark 2.10 It is easy to express the Lamé moduli in terms of the technical moduli:

$$\lambda = \frac{2\nu}{1-2\nu}\,G = \frac{\nu}{(1-2\nu)(1+\nu)}\,E, \quad \mu = G = \frac{1}{2(1+\nu)}\,E.$$

In particular, it follows from these relations that

$$3\lambda + 2\mu = \frac{1}{1-2\nu}\,E.$$

With the use of the technical moduli, the inverse constitutive equation (2.39) reads:

$$\boldsymbol{E} = \frac{1}{E}\Big((1+\nu)\boldsymbol{S} - \nu(\operatorname{tr}\boldsymbol{S})\boldsymbol{I}\Big) = \frac{1}{2G}\Big(\boldsymbol{S} - \frac{\nu}{1+\nu}(\operatorname{tr}\boldsymbol{S})\boldsymbol{I}\Big). \tag{2.45}$$

Consequently, the equal-index components of $\boldsymbol{E}$ are exemplified by

$$E_{11} = \frac{1}{E}\Big(S_{11} - \nu(S_{22} + S_{33})\Big),$$

and the components with different indices by

$$E_{12} = \frac{1}{2G}S_{12},$$

all the other components being obtained via a cyclic permutation of indices. Continuing to use the technical moduli, the direct constitutive equation (2.36) and the strain energy read, respectively,

$$\boldsymbol{S} = \frac{E}{1+\nu}\Big(\boldsymbol{E} + \frac{\nu}{1-2\nu}(\operatorname{tr}\boldsymbol{E})\boldsymbol{I}\Big) = 2G\Big(\boldsymbol{E} + \frac{\nu}{1-2\nu}(\operatorname{tr}\boldsymbol{E})\boldsymbol{I}\Big) \tag{2.46}$$

and

$$\widetilde{\sigma}(\boldsymbol{E}) = \frac{E}{2(1+\nu)}\Big(|\boldsymbol{E}|^2 + \frac{\nu}{1-2\nu}(\operatorname{tr}\boldsymbol{E})^2\Big). \tag{2.47}$$

Remark 2.11 When the extensional rigidity is constant, the differential Eq. (1.6) for the axial deformations of a beam is:

$$w'' + \frac{q}{EA} = 0.$$

It is instructive to demonstrate the mutual consistency of the 3- and 1-D theories of elasticity by 'deducing' (1.6) from Navier equation. This can be done as follows. As in Remark 2.3, restrict attention to displacement fields of the form:

$$\boldsymbol{u}(x) = w(x_3)\boldsymbol{e}_3. \tag{2.48}$$

Then,

$$\Delta \boldsymbol{u} = w''\boldsymbol{e}_3, \quad \operatorname{div} \boldsymbol{u} = w' \quad \Rightarrow \quad \nabla(\operatorname{div} \boldsymbol{u}) = w''\boldsymbol{e}_3,$$

and hence Eq. (2.40) reduces to

$$(\lambda + 2\mu)\, w''\boldsymbol{e}_3 + \boldsymbol{d} = \boldsymbol{0}.$$

At this point, to conclude the announced deduction, it is enough to choose

$$\boldsymbol{d} = \frac{q}{A}\boldsymbol{e}_3$$

and to set

$$\lambda + 2\mu = E. \tag{2.49}$$

It remains for us to convince ourselves that the last position makes sense. Now, it is easy to see that, whenever the strain state

$$\boldsymbol{E} = \tau \boldsymbol{e}_3 \otimes \boldsymbol{e}_3$$

corresponding to a displacement field (2.48) is induced in a linearly elastic isotropic material, the stress state is

$$\boldsymbol{S} = \tau\Big((\lambda + 2\mu)\boldsymbol{e}_3 \otimes \boldsymbol{e}_3 + \lambda(\boldsymbol{e}_1 \otimes \boldsymbol{e}_1 + \boldsymbol{e}_2 \otimes \boldsymbol{e}_2)\Big).$$

Thus, the relative Young'a modulus:

$$E = \frac{\boldsymbol{e}_3 \cdot \boldsymbol{S}\boldsymbol{e}_3}{\boldsymbol{e}_3 \cdot \boldsymbol{E}\boldsymbol{e}_3}$$

has just the expression (2.49).

2.3.3 Plane Stress Fields

A stress field $\boldsymbol{S}$ is said *plane* if there is a Cartesian frame where its representation fulfills a set of conditions formally identical to the conditions (2.19) defining a plane strain field, namely,

$$S_{\alpha\beta} = S_{\alpha\beta}(x_1, x_2), \quad S_{3i} = 0; \tag{2.50}$$

therefore, it has the form:

$$\boldsymbol{S} = S_{11}(x_1, x_2)\boldsymbol{e}_1 \otimes \boldsymbol{e}_1 + S_{12}(x_1, x_2)(\boldsymbol{e}_1 \otimes \boldsymbol{e}_2 + \boldsymbol{e}_2 \otimes \boldsymbol{e}_1) + S_{22}(x_1, x_2)\boldsymbol{e}_2 \otimes \boldsymbol{e}_2.$$

A plane stress field is *balanced for null distance forces* if its divergence is null:

$$\operatorname{div} \boldsymbol{S} = \boldsymbol{0}, \quad (\operatorname{div} \boldsymbol{S})_\alpha = S_{\alpha\beta},_\beta = 0. \tag{2.51}$$

In Sect. 4.2, we shall construct a general representation for those fields $\boldsymbol{S}$ that solve (2.51).

In a linearly elastic isotropic body, a plane stress field induces a strain field that is not plane in general, as an application of the response law (2.45) shows:

$$E_{11} = \frac{1}{E}\Big(S_{11} - \nu S_{22}\Big), \quad E_{22} = \frac{1}{E}\Big(S_{22} - \nu S_{11}\Big), \quad E_{12} = \frac{1}{2G}S_{12},$$
$$E_{3\alpha} = 0, \quad E_{33} = -\frac{\nu}{E}\Big(S_{11} + S_{22}\Big).$$

Quite similarly, a plane strain field does not induce a plane stress field in general, because relations (2.36) and (2.33) imply not only that

$$S_{3\alpha} = 0, \tag{2.52}$$

but also that

$$S_{33} = \lambda(E_{11} + E_{22}) = \frac{\nu E}{(1+\nu)(1-2\nu)}(E_{11} + E_{22}) \neq 0, \text{ in general.}$$

Note for later use that the last relation, when written in terms of stress components and technical moduli, reads:

$$S_{33} = \nu(S_{11} + S_{22}). \tag{2.53}$$

2.3.4 Plane Strain Fields Associated with Plane Stress Fields

Given a plane strain field, it is at times convenient to write the inverse constitutive relation delivering its nonnull components in a fashion formally identical to (2.3.3):

$$E_{11} = \frac{1}{E_0}\Big(S_{11} - \nu_0 S_{22}\Big), \quad E_{22} = \frac{1}{E_0}\Big(S_{22} - \nu_0 S_{11}\Big), \quad E_{12} = \frac{1}{2G}S_{12}, \tag{2.54}$$

where

$$E_0 := \frac{E}{1-\nu^2}, \quad \nu_0 := \frac{\nu}{1-\nu} \tag{2.55}$$

and hence[11]

$$E_0 = 2(1+\nu_0)G. \tag{2.56}$$

A comparison with (2.54) permits to regard the plane strain state (2.3.3) as a part of the strain state induced by a plane stress state in a body made of an isotropic material whose technical moduli are E_0, ν_0, and G.

Given the plane stress $\{S_{11}, S_{22}, S_{12}\}$ and the component S_{33} associated with it by the use of recipe (2.53), the corresponding plane strain is delivered by formulas (2.54)–(2.55). Such a construction is going to be of the essence to solve the 2-D version of Flamant problem with the method we propose.

Remark 2.12 Relation (2.54) can be given a version free from the specialty inherent to the use of Cartesian components and formally identical to $(2.45)_1$:

$$\boldsymbol{E} = \frac{1}{E_0}\Big((1+\nu_0)\boldsymbol{S} - \nu_0(\operatorname{tr}\boldsymbol{S})\boldsymbol{I}_{(2)}\Big), \tag{2.57}$$

where $\boldsymbol{S}$ is, as anticipated, a plane stress field and $\boldsymbol{I}_{(2)}$ denotes the two-dimensional identity tensor.

[11] For example, let us show how the first of (2.54) is arrived at: from $(2.33)_{1,2}$ we have that

$$S_{11} = \frac{E}{1+\nu}\Big(E_{11} + \frac{\nu}{1-2\nu}(E_{11}+E_{22})\Big), \quad S_{11}+S_{22} = \frac{E}{(1+\nu)(1-2\nu)}(E_{11}+E_{22});$$

consequently,

$$E_{11} = \frac{1+\nu}{E}S_{11} - \frac{\nu}{1-2\nu}\frac{(1+\nu)(1-2\nu)}{E}(S_{11}+S_{22}) = \frac{1+\nu}{E}\big(S_{11} - \nu(S_{11}+S_{22})\big) \text{ etc.}$$

2.4 Compatibility in Stress Terms

2.4.1 The Three-Dimensional Case

When the response mapping of a linearly elastic material is invertible, the compatibility condition (2.12) is written in stress terms in a straightforward manner, with the use of (2.35):

$$\operatorname{curl}\operatorname{curl}(\mathbb{C}^{-1}\boldsymbol{S}) = \boldsymbol{0}. \tag{2.58}$$

When an equilibrium problem is formulated in stress terms, the symmetric-valued fields $\boldsymbol{S}$ to be inserted in (2.58) must satisfy the equilibrium equation (2.31). In the applications we are interested in, three conditions hold, which make special and easy to handle the compatibility condition (2.58):

(i) the material is supposed to be isotropic, and hence, in view of $(2.45)_2$,

$$\mathbb{C}^{-1}\boldsymbol{S} = \frac{1}{2G}\Big(\boldsymbol{S} - \frac{\nu}{1+\nu}(\operatorname{tr}\boldsymbol{S})\boldsymbol{I}\Big); \tag{2.59}$$

(ii) the bodies under examination are supposed homogeneous, hence the elastic moduli are spatially constant;

(iii) distance actions are supposed to be null, and hence

$$\operatorname{div}\boldsymbol{S} = \boldsymbol{0}. \tag{2.60}$$

We now proceed to determine the form of condition (2.58) under these circumstances.

Firstly, it follows from (2.58), (2.59), and assumption (ii), that

$$\operatorname{curl}\operatorname{curl}\boldsymbol{S} - \frac{\nu}{1+\nu}\operatorname{curl}\operatorname{curl}\big((\operatorname{tr}\boldsymbol{S})\boldsymbol{I}\big) = \boldsymbol{0}. \tag{2.61}$$

To move further, we observe that each smooth symmetric-valued tensor field $\boldsymbol{A}$ satisfies identically the differential condition:

$$\begin{aligned}\operatorname{curl}\operatorname{curl}\boldsymbol{A} &= -\Delta\boldsymbol{A} - \nabla(\nabla(\operatorname{tr}\boldsymbol{A})) + \nabla(\operatorname{div}\boldsymbol{A})\\ &\quad + (\nabla(\operatorname{div}\boldsymbol{A}))^T + (\Delta(\operatorname{tr}\boldsymbol{A}) - \operatorname{div}(\operatorname{div}\boldsymbol{A}))\boldsymbol{I}\end{aligned} \tag{2.62}$$

(cf. [6], Sect. 14); in components,

$$e_{ijk}e_{lmn}A_{jm},_{kn} = -A_{il},_{jj} - A_{jj},_{il} + A_{ij},_{jl} + A_{lj},_{ji} + (A_{jj},_{kk} - A_{jk},_{jk})\delta_{il}. \tag{2.63}$$

Consequently,

$$\operatorname{tr}\big(\operatorname{curl}\operatorname{curl}\boldsymbol{A}\big) = \Delta(\operatorname{tr}\boldsymbol{A}) - \operatorname{div}(\operatorname{div}\boldsymbol{A}); \tag{2.64}$$

moreover, for $\boldsymbol{A} = \alpha\boldsymbol{I}$, (2.62) yields:

$$\operatorname{curl}\operatorname{curl}\big(\alpha\boldsymbol{I}\big) = (\Delta\alpha)\boldsymbol{I} - \nabla(\nabla\alpha), \tag{2.65}$$

hence, in particular,

$$\operatorname{tr}\big(\operatorname{curl}\operatorname{curl}\big(\alpha\boldsymbol{I}\big)\big) = 2\Delta\alpha. \tag{2.66}$$

Thus, if a field $\boldsymbol{S}$ satisfying (2.60) is compatible, then necessarily it must be such that

$$\Delta(\operatorname{tr}\boldsymbol{S}) = 0, \tag{2.67}$$

a relation that is arrived at by taking the trace of (2.61), with the use of (2.64) and (2.66) and of the constitutive inequalities restricting the admissible values of ν (recall Remark 2.7). Due to this partial result, we deduce from (2.62) that

$$\operatorname{curl}\operatorname{curl}\boldsymbol{S} = -\Delta\boldsymbol{S} - \nabla\nabla(\operatorname{tr}\boldsymbol{S}),$$

and from (2.65) that

$$\operatorname{curl}\operatorname{curl}\big((\operatorname{tr}\boldsymbol{S})\boldsymbol{I}\big) = -\nabla\nabla(\operatorname{tr}\boldsymbol{S});$$

On taking the two last relations into account, (2.61) becomes the sought-for *compatibility condition in stress terms*:

$$\Delta\boldsymbol{S} + \frac{1}{1+\nu}\nabla\nabla(\operatorname{tr}\boldsymbol{S}) = \boldsymbol{0}. \tag{2.68}$$

Remark 2.13 Once a general representation has been found for all solutions of the equilibrium equation (2.60), we are going to use condition (2.68) to select those associable with strain and stress fields consistent with the constitutive behavior of the material under consideration. Remarkably, this behavior affects (2.68) only through the Poisson's modulus. A *universal* stress field—that is, a stress field being balanced and compatible for whatever isotropic material—must satisfy, in addition to (2.60), a system even more stringent than (2.68), namely,

$$\Delta\boldsymbol{S} = \boldsymbol{0}, \quad \nabla\nabla(\operatorname{tr}\boldsymbol{S}) = \boldsymbol{0}.$$

2.4.2 The Two-Dimensional Case

An assigned plane strain field whose Cartesian components are E_{11}, E_{22}, E_{12} is compatible if condition (2.20) holds; we repeat it here for the reader's convenience:

$$2\,E_{12,12} = E_{11,22} + E_{22,11}.$$

This condition can be written in terms of stresses with the use of the constitutive relations (2.54)–(2.56). One begins by finding:

$$\frac{1}{G}S_{12,12} = \frac{1}{E_0}\Big(S_{11,22} + S_{22,11} - \nu_0\big(S_{11,11} + S_{22,22}\big)\Big),$$

a relation that can be given the intermediate form

$$S_{12},_{12} = \frac{1}{2(1 + \nu_0)}\Big((S_{11} + S_{22}),_{11} + (S_{11} + S_{22}),_{22} - (1 + \nu_o)(S_{11},_{11} + S_{22},_{22})\Big),$$

and then the final form

$$S_{\alpha\alpha},_{\beta\beta} = (1 + \nu_0)S_{\alpha\beta},_{\alpha\beta}.$$

When the field $\boldsymbol{S}$ is plane, the last condition can be witten more compactly:

$$\Delta(\operatorname{tr}\boldsymbol{S}) = (1 + \nu_0)\operatorname{div}(\operatorname{div}\boldsymbol{S}).$$

A consequence of this result, of paramount importance in certain developments to come, is the condition that a plane stress field, balanced for null distance forces, must satisfy to be compatible. In view of (2.50) and (2.51), that condition is:

$$\Delta(\operatorname{tr}\boldsymbol{S}) = 0. \tag{2.69}$$

It is not difficult to check that the same condition guarantees the compatibility of the three-dimensional stress field

$$\widetilde{\boldsymbol{S}} = \boldsymbol{S} + S_{33}\boldsymbol{e}_3 \otimes \boldsymbol{e}_3, \quad S_{33} = \nu S_{\alpha\alpha}. \tag{2.70}$$

Such a stress field, by way of the constitutive relations (2.54), is associable with a compatible plane strain field, which in turn is associated with a plane displacement field.

References

1. Capriz G, Podio-Guidugli P (2004) Whence the boundary conditions in modern continuum physics? In: Capriz G, Podio-Guidugli P (eds) Atti dei Convegni Lincei N. 210. Accademia dei Lincei, Roma, pp 19–42.
2. Carillo S, Podio-Guidugli P, Vergara Caffarelli G (2002) Second-order surface potentials in finite elasticity. In: Brocato M, Podio-Guidugli P (eds) Rational Continua. Classical and New, Springer Italia, Milano, pp 19–38
3. Cauchy AL (1823) Recherches sur l'équilibre et le mouvement intérieur des corps solides ou fluides, élastiques ou non élastiques. Bull Soc Philom Paris 9–13:300–304
4. DiCarlo A (2004) Actual surfaces versus virtual cuts. In: Capriz G, Podio-Guidugli P (eds) Atti dei Convegni Lincei N. 210. Accademia dei Lincei, Roma, pp 97–113.
5. Gladwell GML (1980) Contact problems in the classsical theory of elasticity. Sijthoff and Noordhoff, Alphen aan den Rijn
6. Gurtin ME (1972) The Linear theory of elasticity. In: Flügge S (ed) Handbuch der Physik, vol. 2. Springer, Berlin

7. Gurtin ME (1981) An introduction to continuum mechanics. Academic Press, Boston
8. Love AEH (1927) A treatise on the mathematical theory of elasticity. Cambridge University Press, Cambridge
9. Malvern LE (1969) Introduction to the mechanics of a continuous medium. Prentice-Hall, Englewood Cliffs
10. Meriam JL, Kraige LG (2003) Engineering mechanics, vol 1. Wiley, New York
11. Podio-Guidugli P (2000) A primer in elasticity. Kluwer, Dordrecht
12. Podio-Guidugli P (2004) Examples of concentrated contact interactions in simple bodies. J Elasticity 75:167–186
13. Podio-Guidugli P (2005) On concentrated interactions. In: Dal Maso G, De Simone A, Tomarelli F (eds) Variational Problems in Materials Science. Birkhäuser, Basel
14. Podio-Guidugli P, Schuricht F (2012) Concentrated actions on plane cuspidate bodies. J Elasticity 106:107–114
15. Schuricht F (2007) A new mathematical foundation for contact interactions in continuum physics. Arch Rat Mech Anal 184:495–551
16. Schuricht F (2007) Interactions in continuum physics. In: Silhavý M (ed) Mathematical modeling of bodies with complicated bulk and boundary behavior. Quaderni di Matematica 20:169–196
17. Sokolnikoff IS (1956) Mathematical theory of elasticity. McGraw-Hill, New York

Chapter 3
Geometric and Analytic Tools

The problems in classical elasticity we tackle have intrinsic symmetries that are best exploited with the use of ad hoc coordinate systems, because the associated vector and tensor bases allow for convenient representations of the fields of interest and their transformations under the action of differential operators. In this chapter we collect a *modicum* of basic material from differential geometry and analysis.

3.1 Curvilinear Coordinates, Covariant and Contravariant Bases

Let x any chosen point in $\mathcal{E}^N$, whose position vector with respect to a chosen origin $o \in \mathcal{E}^N$ is, we recall, $\boldsymbol{x} = x - o$. Alongside with its Cartesian coordinates x_i, we associate to x one or more sets of curvilinear coordinates ζ^i, that is, of ordered N-tuples of real numbers, each of which is such that the following mappings are all bijective:

$$(\zeta^1, \ldots, \zeta^N) \leftrightarrow x \leftrightarrow \boldsymbol{x} \leftrightarrow (x_1, \ldots, x_N).$$

Two sets of N *basis vectors* are defined at each point $x \in \mathcal{E}^N$, those composing the *covariant basis*:

$$\boldsymbol{g}_i := \partial_{\zeta_i} x = \partial_{\zeta_i} \boldsymbol{x} \tag{3.1}$$

and those composing the *contravariant basis*:

$$\boldsymbol{g}^i := \nabla \zeta^i; \tag{3.2}$$

while neither type of basis is in general orthogonal or consists of unit vectors, it follows from definitions (3.1) and (3.2) that

$$\boldsymbol{g}^i \cdot \boldsymbol{g}_k = \delta_{ik}.$$

P. Podio-Guidugli and A. Favata, *Elasticity for Geotechnicians*, Solid Mechanics and Its Applications 204, DOI: 10.1007/978-3-319-01258-2_3,

Two equivalent dyadic representations in terms of basis vectors can be given for the *identity tensor* $\boldsymbol{I}$, namely,

$$\boldsymbol{I} = \boldsymbol{g}_i \otimes \boldsymbol{g}^i = \boldsymbol{g}^i \otimes \boldsymbol{g}_i .$$

For $\boldsymbol{v}$ any vector, this relation implies that

$$\boldsymbol{v} = \boldsymbol{I}\boldsymbol{v} = (\boldsymbol{g}^i \cdot \boldsymbol{v})\boldsymbol{g}_i = (\boldsymbol{g}_i \cdot \boldsymbol{v})\boldsymbol{g}^i ; \tag{3.3}$$

the *covariant* and *contravariant components* of v are, respectively,

$$v_i := \boldsymbol{v} \cdot \boldsymbol{g}_i \quad \text{and} \quad v^i := \boldsymbol{v} \cdot \boldsymbol{g}^i ,$$

whence

$$\boldsymbol{v} = v^i \boldsymbol{g}_i = v_i \boldsymbol{g}^i . \tag{3.4}$$

Now, the operation of taking the Cartesian components of a vector has no effect on physical dimensions, because all vectors of a Cartesian basis are dimensionless. Instead, when general curvilinear coordinates are used, the physical dimensions of v^i and v_i may be neither all the same nor the same as $\boldsymbol{v}$, because one or another of the basis vectors may have non-zero dimensions.[1] In applications, this fact may blur the physical perception of a quantity undergoing algebraic or differential manipulations. In the next section we show how this potential difficulty is removed.

3.2 Orthogonal Coordinates, Physical Bases

What type of curvilinear coordinates to use is suggested by the symmetries inherent to the problem at hand. We begin by the simplest instance of polar coordinates, to be used in the study of the plane Boussinesq-Flamant Problem (Sect. 4.3).

3.2.1 Polar Coordinates

Let $N = 2$, and choose

$$\zeta^1 = \rho, \quad \zeta^2 = \vartheta$$

(Fig. 3.1).

---

[1] It follows from (3.1) and (3.2) that, for each index i, vectors $\boldsymbol{g}_i$ and $\boldsymbol{g}^i$ have inverse physical dimensions:

$$\dim(\boldsymbol{g}_i) = \big(\dim(\boldsymbol{g}^i)\big)^{-1}.$$

Thus, in particular, the identity tensor is dimensionless, as required implicitly by $(3.3)_1$; moreover, in view of (3.4),

$$\dim(v_i) = \big(\dim(v^i)\big)^{-1}.$$

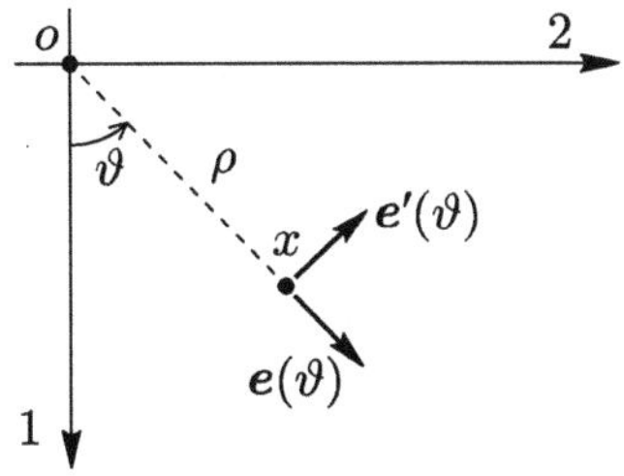

Fig. 3.1 The Flamant half plane: coordinates, basis vectors and applied load

The position vector of a point $x \in \mathcal{E}^2$ is

$$\boldsymbol{x} = x_\alpha\, \boldsymbol{e}_\alpha = \rho\widehat{\boldsymbol{e}}(\vartheta),$$

where

$$x_1 = \rho\cos\vartheta, \quad x_2 = \rho\sin\vartheta,$$

so that, in particular,

$$\widehat{\boldsymbol{e}}(\vartheta) = \cos\vartheta\boldsymbol{e}_1 + \sin\vartheta\boldsymbol{e}_2;$$

and where

$$\rho^2 = x_1^2 + x_2^2 = \boldsymbol{x}\cdot\boldsymbol{x}, \quad \tan\vartheta = \frac{x_2}{x_1} = \frac{\boldsymbol{x}\cdot\boldsymbol{e}_2}{\boldsymbol{x}\cdot\boldsymbol{e}_1}. \tag{3.5}$$

On applying (3.1) and (3.2), we find:

$$\begin{aligned} \boldsymbol{g}_1 &= \partial_\rho x = \widehat{\boldsymbol{e}}(\vartheta), \\ \boldsymbol{g}_2 &= \partial_\vartheta x = \rho\widehat{\boldsymbol{e}}'(\vartheta), \end{aligned}$$

for the covariant basis, and

$$\begin{aligned} \boldsymbol{g}^1 &= \nabla\rho = \frac{\partial\rho}{\partial x_\alpha}\boldsymbol{e}_\alpha = \widehat{\boldsymbol{e}}(\vartheta), \\ \boldsymbol{g}^2 &= \nabla\vartheta = \frac{\partial\vartheta}{\partial x_\alpha}\boldsymbol{e}_\alpha = \rho^{-1}\widehat{\boldsymbol{e}}'(\vartheta), \end{aligned} \tag{3.6}$$

for the contravariant basis; consequently, the 2-D identity tensor has the following representations[2]:

[2] On differentiating the first of (3.5), we find that

$$2\rho\nabla\rho = 2\boldsymbol{x} \quad\Rightarrow\quad \nabla\rho = \rho^{-1}\boldsymbol{x}, \text{ with } \rho = |\boldsymbol{x}|;$$

on differentiating the second, that

$$\frac{1}{\cos^2\vartheta}\nabla\vartheta = \frac{(\boldsymbol{x}\cdot\boldsymbol{e}_1)\boldsymbol{e}_2 - (\boldsymbol{x}\cdot\boldsymbol{e}_2)\boldsymbol{e}_1}{(\boldsymbol{x}\cdot\boldsymbol{e}_1)^2}, \text{ con } \cos\vartheta = \rho^{-1}(\boldsymbol{x}\cdot\boldsymbol{e}_1), \ \sin\vartheta = \rho^{-1}(\boldsymbol{x}\cdot\boldsymbol{e}_2),$$

whence $(3.6)_2$.

$$\boldsymbol{I}_{(2)} = \boldsymbol{g}_\alpha \otimes \boldsymbol{g}^\alpha = \boldsymbol{g}^\alpha \otimes \boldsymbol{g}_\alpha = \boldsymbol{e} \otimes \boldsymbol{e} + \boldsymbol{e}' \otimes \boldsymbol{e}'.$$

Remark 3.1 The case of polar coordinates exemplifies the general situation, where, with the exception of Cartesian coordinates, covariant and contravariant bases both differ and depend on the point. In a way, the price to pay for the advantage of representing vector and tensor fields in geometrically convenient bases is that those bases must be adjourned according to the point in space where the representation is sought: usually, the game is worth the candle.

Polar coordinates are said *orthogonal*, because such are the vectors of each of the two bases. The basis vectors $\boldsymbol{g}_1$ and $\boldsymbol{g}^1$ have unit length and are dimensionless; not so the vectors $\boldsymbol{g}_2$ and $\boldsymbol{g}^2$, because $\dim(\boldsymbol{g}_2) = L$ and hence, $\dim(\boldsymbol{g}^2) = L^{-1}$ (recall the contents of the footnote at the end of Sect. 3.1). To avoid representations where the dimensions of all components are not the same, in this and in all cases when the coordinate system at hand is orthogonal, one customarily introduces the *physical basis*

$$\boldsymbol{g}_{<i>} := |\boldsymbol{g}_i|^{-1}\boldsymbol{g}_i = |\boldsymbol{g}^i|^{-1}\boldsymbol{g}^i \quad \text{(index } i \text{ unsummed)}.$$

Accordingly, for polar coordinates the physical basis is

$$\boldsymbol{g}_{<1>} = \widehat{\boldsymbol{e}}'(\vartheta), \quad \boldsymbol{g}_{<2>} = \widehat{\boldsymbol{e}}(\vartheta)$$

In this book, *we shall employ exclusively orthogonal curvilinear coordinates and physical bases*. Therefore, neither superscript indices nor the symbol $\langle\cdot\rangle$ will be needed: components will be distinguished only by subscript indices.

3.2.2 Cylindrical Coordinates

For $N = 3$, we consider two types of *cylindrical coordinates*:

(i) $\zeta^1 = \rho, \quad \zeta^2 = \vartheta, \quad \zeta^3 = x_3$ and

(ii) $\zeta^1 = z, \quad \zeta^2 = r, \quad \zeta^3 = \varphi,$

with

$$z^2 + r^2 = \rho^2, \quad z^{-1}r = \tan\vartheta, \quad \sin\varphi = \frac{x_3}{\rho\sin\vartheta}$$

(see Fig. 3.2). We make use of coordinates of the former type for the Flamant Problem (Chap. 4), of the latter for the Boussinesq Problem (Chap. 5).

With reference to Fig. 3.2, in terms of the cylindrical coordinates z, r, φ, the Cartesian coordinates of a point x with respect to the origin o are:

$$x_1 = z, \quad x_2 = r\cos\varphi, \quad x_3 = r\sin\varphi.$$

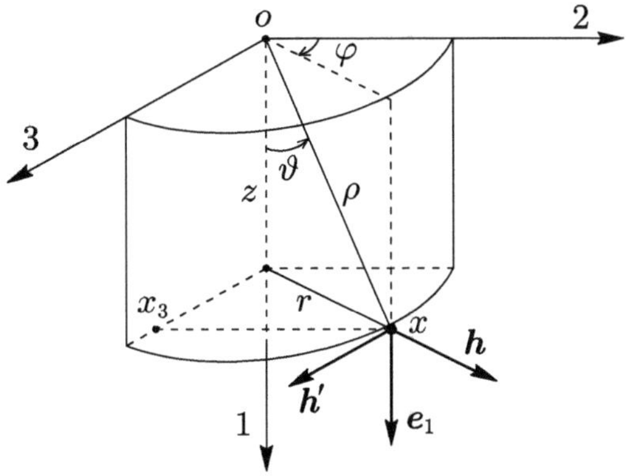

Fig. 3.2 The cylindrical coordinates (ρ, ϑ, x_3) and (z, r, φ)

Consequently, the position vector of x with respect to o is:

$$\boldsymbol{x} = z\boldsymbol{e}_1 + r\widehat{\boldsymbol{h}}(\varphi), \quad \text{with } z, r \geq 0, \ \ \varphi \in [0, 2\pi),$$

where

$$\widehat{\boldsymbol{h}}(\varphi) = \cos\varphi\, \boldsymbol{e}_2 + \sin\varphi\, \boldsymbol{e}_3.$$

The mutually orthogonal *contravariant* basis vectors are

$$\begin{aligned} \boldsymbol{g}^1 &= \boldsymbol{e}_1, \\ \boldsymbol{g}^2 &= \boldsymbol{h}, \\ \boldsymbol{g}^3 &= r^{-1}\boldsymbol{h}'; \end{aligned} \tag{3.7}$$

the *covariant* basis vectors are:

$$\begin{aligned} \boldsymbol{g}_1 &= \boldsymbol{e}_1, \\ \boldsymbol{g}_2 &= \boldsymbol{h}, \\ \boldsymbol{g}_3 &= r\boldsymbol{h}'; \end{aligned}$$

the physical basis is the following triplet of unit vectors: $(\boldsymbol{e}_1, \boldsymbol{h}, \boldsymbol{h}')$.[3] The associated physical basis for the space of symmetric tensors is the following sextuple of dyads:

[3] To derive (3.7), recall that:

$$\boldsymbol{g}^1 := \nabla z, \quad z = \boldsymbol{x}\cdot\boldsymbol{e}_1; \qquad \boldsymbol{g}^2 := \nabla r, \quad r^2 = (\boldsymbol{x}\cdot\boldsymbol{e}_2)^2 + (\boldsymbol{x}\cdot\boldsymbol{e}_3)^2; \qquad \boldsymbol{g}^3 := \nabla\varphi, \quad \tan\varphi = \frac{\boldsymbol{x}\cdot\boldsymbol{e}_3}{\boldsymbol{x}\cdot\boldsymbol{e}_2}.$$

$$
\begin{aligned}
&\boldsymbol{e}_1 \otimes \boldsymbol{e}_1, \;\; \boldsymbol{h} \otimes \boldsymbol{h}, \;\; \boldsymbol{h}' \otimes \boldsymbol{h}', \\
&\frac{1}{\sqrt{2}}(\boldsymbol{e}_1 \otimes \boldsymbol{h} + \boldsymbol{h} \otimes \boldsymbol{e}_1), \\
&\frac{1}{\sqrt{2}}(\boldsymbol{e}_1 \otimes \boldsymbol{h}' + \boldsymbol{h}' \otimes \boldsymbol{e}_1), \\
&\frac{1}{\sqrt{2}}(\boldsymbol{h} \otimes \boldsymbol{h}' + \boldsymbol{h}' \otimes \boldsymbol{h});
\end{aligned}
\tag{3.8}
$$

in particular, the identity tensor has the representation:

$$
\boldsymbol{I} = \boldsymbol{e}_1 \otimes \boldsymbol{e}_1 + \boldsymbol{h} \otimes \boldsymbol{h} + \boldsymbol{h}' \otimes \boldsymbol{h}'.
$$

3.2.3 Spherical Coordinates

Spherical coordinates:

$$
\zeta^1 = \rho, \quad \zeta^2 = \vartheta, \quad \zeta^3 = \varphi, \tag{3.9}
$$

with

$$
(\rho, \vartheta, \varphi) \in [0, +\infty) \times [-\pi, +\pi) \times [0, \pi),
$$

(see Fig. 3.3) are best for the Kelvin Problem (Sect. 6.3).

In terms of spherical coordinates, the Cartesian coordinates of a point x with respect to the origin o are:

$$
x_1 = \rho \cos\vartheta, \quad x_2 = \rho \sin\vartheta \cos\varphi, \quad x_3 = \rho \sin\vartheta \sin\varphi; \tag{3.10}
$$

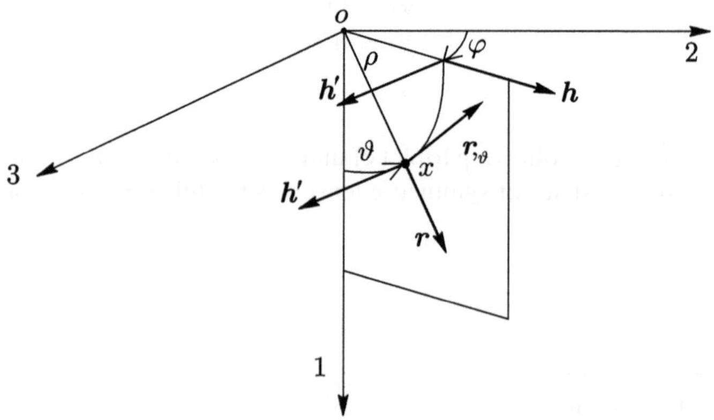

Fig. 3.3 The spherical coordinates $(\rho, \vartheta, \varphi)$

the reader is advised to check that, for $\varphi = 0$, these formulas (as well as other in this subsection) reduce to the corresponding formulas for polar coordinates. It follows from (3.10) that the position vector of x is:

$$\boldsymbol{x} = \rho\widehat{\boldsymbol{r}}(\vartheta, \varphi), \tag{3.11}$$

where

$$\boldsymbol{r} := \widehat{\boldsymbol{r}}(\vartheta, \varphi) = \cos\vartheta\, \boldsymbol{e}_1 + \sin\vartheta\, \widehat{\boldsymbol{h}}(\varphi), \quad \boldsymbol{h} := \widehat{\boldsymbol{h}}(\varphi) = \cos\varphi\, \boldsymbol{e}_2 + \sin\varphi\, \boldsymbol{e}_3.$$

The *contravariant* basis vectors are[4]:

$$\begin{aligned} \boldsymbol{g}^1 &= \boldsymbol{r}, \\ \boldsymbol{g}^2 &= \rho^{-1}(-\sin\vartheta \boldsymbol{e}_1 + \cos\vartheta \boldsymbol{h}) = \rho^{-1}\boldsymbol{r},_\vartheta\,, \\ \boldsymbol{g}^3 &= (\rho\sin\vartheta)^{-1}\boldsymbol{h}' = (\rho\sin^2\vartheta)^{-1}\boldsymbol{r},_\varphi\,. \end{aligned} \tag{3.12}$$

It is easy to check that these vectors are mutually orthogonal, as are the *covariant* basis vectors:

$$\begin{aligned} \boldsymbol{g}_1 &= \boldsymbol{x},_\rho = \boldsymbol{r}, \\ \boldsymbol{g}_2 &= \boldsymbol{x},_\vartheta = \rho\boldsymbol{r},_\vartheta = \rho(-\sin\vartheta\boldsymbol{e}_1 + \cos\vartheta\boldsymbol{h}), \\ \boldsymbol{g}_3 &= \boldsymbol{x},_\varphi = \rho\boldsymbol{r},_\varphi = (\rho\sin\vartheta)\boldsymbol{h}'; \end{aligned}$$

with this, one finds that

$$\boldsymbol{I} = \boldsymbol{g}^i \otimes \boldsymbol{g}_i = \boldsymbol{g}_j \otimes \boldsymbol{g}^j = \boldsymbol{r} \otimes \boldsymbol{r} + \boldsymbol{r},_\vartheta \otimes \boldsymbol{r},_\vartheta + \boldsymbol{h}' \otimes \boldsymbol{h}',$$

a representation of the identity tensor where the vectors $(\boldsymbol{r}, \boldsymbol{r},\ _\vartheta, \boldsymbol{h}')$ of the physical basis appear.

Remark 3.2 Note the following relationships between the cylindrical and spherical coordinates of a given point:

$$z = \rho\cos\vartheta, \quad r = \rho\,|\sin\vartheta|; \quad \rho^2 = z^2 + r^2, \quad |\tan\vartheta| = \frac{r}{z}.$$

---

[4] To derive (3.12), recall that:

$$\boldsymbol{g}^1 = \nabla\rho, \quad \nabla(\boldsymbol{x}\cdot\boldsymbol{x}) = \nabla(\rho^2), \quad \rho = |\boldsymbol{x}|;$$

$$\boldsymbol{g}^2 = \nabla\vartheta, \quad \boldsymbol{e}_1 = \nabla(\boldsymbol{x}\cdot\boldsymbol{e}_1) = \nabla(\rho\cos\vartheta) = (\cos\vartheta)\nabla\rho + \rho\nabla(\cos\vartheta) = \cos\vartheta\,\boldsymbol{r} - \rho\sin\vartheta\nabla\vartheta;$$

$$\boldsymbol{g}^3 = \nabla\varphi, \quad \frac{\cos\varphi\boldsymbol{e}_3 - \sin\varphi\boldsymbol{e}_2}{\rho\sin\vartheta\cos^2\varphi} = \frac{(\boldsymbol{x}\cdot\boldsymbol{e}_2)\boldsymbol{e}_3 - (\boldsymbol{x}\cdot\boldsymbol{e}_3)\boldsymbol{e}_2}{(\boldsymbol{x}\cdot\boldsymbol{e}_2)^2} = \nabla\Big(\frac{\boldsymbol{x}\cdot\boldsymbol{e}_3}{\boldsymbol{x}\cdot\boldsymbol{e}_2}\Big) = \nabla(\tan\varphi) = \frac{1}{\cos^2\varphi}\nabla\varphi.$$

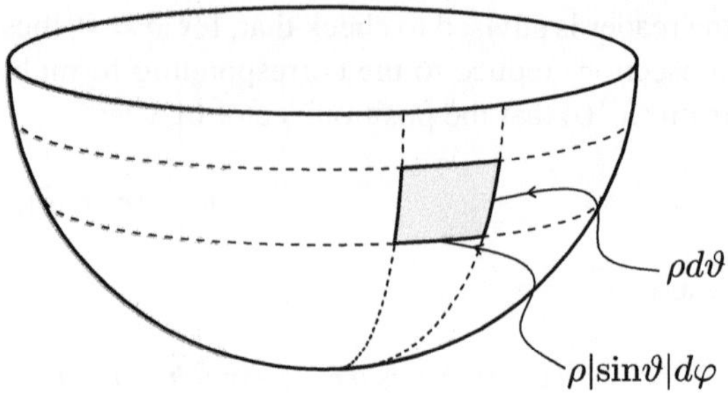

Fig. 3.4 The area element of the coordinate surface S_ρ

Remark 3.3 The outer normal $\boldsymbol{n} = \boldsymbol{r},_\vartheta \times \boldsymbol{h}'$ to any coordinate surface $\mathcal{S}_\rho := \{x \mid |x - o| = \rho\}$ coincides with the unit vector $\boldsymbol{r}$; the oriented area element of such a surface has the expression:

$$\boldsymbol{n}(\vartheta, \varphi)da = \rho^2 \boldsymbol{r}(\vartheta, \varphi)|\sin\vartheta|\, d\vartheta d\varphi$$

(Fig. 3.4). Needless to say, the area of $\mathcal{S}_\rho$ is

$$\rho^2 \int_0^\pi \left(\int_{-\pi}^{+\pi} |\sin\vartheta|\, d\vartheta \right) d\varphi = 4\pi\rho^2;$$

and, as is intuitively true, the average of the normal field over $\mathcal{S}_\rho$ is null, because

$$\rho^{-2} \int_{\mathcal{S}_\rho} \boldsymbol{n}(\vartheta, \varphi)da = a\,\boldsymbol{e}_1 + b\,\boldsymbol{e}_3, \quad \int_0^\pi \boldsymbol{h}(\varphi)d\varphi = 2\boldsymbol{e}_3,$$

$$a := \pi \int_{-\pi}^{+\pi} \cos\vartheta\, |\sin\vartheta|\, d\vartheta = 0, \quad b := 2 \int_{-\pi}^{+\pi} \sin\,\vartheta\, |\sin\vartheta|\, d\vartheta = 0.$$

By the same token, we find that

$$\rho^{-2} \int_{\frac{1}{2}\mathcal{S}_\rho} \boldsymbol{n}(\vartheta, \varphi)da = \left(\pi \int_{-\pi/2}^{+\pi/2} \cos\vartheta\, |\sin\vartheta|\, d\vartheta \right) \boldsymbol{e}_1 = \pi\,\boldsymbol{e}_1,$$

where $\frac{1}{2}\mathcal{S}_\rho$ denotes the half of $\mathcal{S}_\rho$ below the coordinate plane $x_1 = 0$. It will be important for certain developments to come to realize that, whenever a given function f is even,

$$\int_{\frac{1}{2}\mathcal{S}_\rho} f(\vartheta)\boldsymbol{n}(\vartheta, \varphi)da = \pi \left(\int_{-\pi/2}^{+\pi/2} f(\vartheta) \cos\vartheta\, |\sin\vartheta|\, d\vartheta \right) \boldsymbol{e}_1, \tag{3.13}$$

whereas

$$\int_{\frac{1}{2}S_\rho} f(\vartheta)\boldsymbol{n}(\vartheta,\varphi)da = 2\left(\int_{-\pi/2}^{+\pi/2} f(\vartheta)\sin\vartheta\,|\sin\vartheta|\,d\vartheta\right)\boldsymbol{e}_3$$

whenever f is odd.

3.3 Representations of Differential Operators

Differential operators have coordinate-dependent representations, but one and only one intrinsic definition. Conceptually, only the latter counts, but operationally it is expedient to know, problem by problem, the most effective representation of the former type.

In our developments to come, the first differential operator to be introduced is the *gradient of a scalar field* $\varphi = \widetilde{\varphi}(x)$, whose intrinsic definition is:

$$\nabla\widetilde{\varphi}(x)\cdot\boldsymbol{h} := \lim_{\varepsilon\to 0}\frac{\widetilde{\varphi}(x+\varepsilon\boldsymbol{h}) - \widetilde{\varphi}(x)}{\varepsilon}\,. \tag{3.14}$$

This definition delivers a vector field, which has the following Cartesian representation:

$$\nabla\breve{\varphi}(x_1,x_2,x_3) = \breve{\varphi},_i(x_1,x_2,x_3)\boldsymbol{e}_i\,,$$

where

$$\breve{\varphi}(x_1,x_2,x_3) := \widetilde{\varphi}(x),\quad x = \breve{x}(x_1,x_2,x_3).$$

For a generic system of curvilinear coordinates, one sets:

$$\widehat{\varphi}(\zeta^1,\zeta^2,\zeta^3) := \widetilde{\varphi}(x),\quad \zeta^i = \widetilde{\zeta}^i(x),$$

whence the representation

$$\nabla\widehat{\varphi}(\zeta^1,\zeta^2,\zeta^3) = \widehat{\varphi},_{\zeta^i}(\zeta^1,\zeta^2,\zeta^3)\nabla\zeta^i(x),\quad x = \widehat{x}(\zeta^1,\zeta^2,\zeta^3);$$

in short, in view of definition (3.2),

$$\nabla\varphi := \varphi,_i\,\boldsymbol{g}^i\,. \tag{3.15}$$

where vectors $\boldsymbol{g}^i$ compose the contravariant basis.

On exploiting the notion of gradient of a scalar field, one can easily build the definition of gradient for a vector field $\boldsymbol{v}$: it is enough to choose $\varphi = \boldsymbol{v}\cdot\boldsymbol{a}$ in (3.14), with $\boldsymbol{a}$ an arbitrary vector; then, the *gradient of a vector field* $\boldsymbol{v}$ is:

$$\nabla\boldsymbol{v} = \boldsymbol{v},_i\otimes\boldsymbol{g}^i\,. \tag{3.16}$$

With the use of this relation, the *divergence of a vector field* $\boldsymbol{v}$ is defined to be:

$$\operatorname{div} \boldsymbol{v} := \operatorname{tr}(\nabla \boldsymbol{v}) = \nabla \boldsymbol{v} \cdot \boldsymbol{I} = \boldsymbol{v}_{,i} \cdot \boldsymbol{g}^i .$$

In turn, when the last definition is used for the vector field $\boldsymbol{v} = \boldsymbol{V}^T \boldsymbol{a}$ (with $\boldsymbol{a}$, as usual, an arbitrary constant vector), the definition of *divergence of a tensor field* $\boldsymbol{V}$ is:

$$\operatorname{div} \boldsymbol{V} = \boldsymbol{V}_{,i} \boldsymbol{g}^i . \tag{3.17}$$

Remark 3.4 For polar coordinates, (3.15) becomes:

$$\nabla \varphi = \varphi_{,\alpha}\, \boldsymbol{g}^\alpha = \varphi_{,\rho}\, \boldsymbol{e} + \rho^{-1} \varphi_{,\vartheta} \boldsymbol{e}' ,$$

while (3.17) becomes:

$$\operatorname{div} \boldsymbol{V} = \boldsymbol{V}_{,\alpha}\, \boldsymbol{g}^\alpha = \boldsymbol{V}_{,\rho}\, \boldsymbol{e} + \rho^{-1} \boldsymbol{V}_{,\vartheta}\, \boldsymbol{e}' . \tag{3.18}$$

Lastly, by posing $\boldsymbol{v} = \nabla \varphi$ in (3.16), it is not difficult to deduce from that relation that

$$\begin{aligned} \nabla^{(2)} \varphi &= (\nabla \varphi)_{,\alpha} \otimes \boldsymbol{g}^\alpha = \varphi_{,\rho\rho}\, \boldsymbol{e} \otimes \boldsymbol{e} + \rho^{-1} (\varphi_{,\rho} + \rho^{-1}\, \varphi_{,\vartheta\vartheta}) \boldsymbol{e}' \otimes \boldsymbol{e}' \\ &\quad + \rho^{-1} (\varphi_{,\rho\vartheta} - \rho^{-1}\, \varphi_{,\vartheta}) (\boldsymbol{e} \otimes \boldsymbol{e}' + \boldsymbol{e}' \otimes \mathbf{e}), \\ \Delta \varphi &= \operatorname{tr}(\nabla^{(2)} \varphi) = \varphi_{,\rho\rho} + \rho^{-1}\, \varphi_{,\rho} + \rho^{-2}\, \varphi_{,\vartheta\vartheta} . \end{aligned} \tag{3.19}$$

Part II
Three Classical Problems: Flamant's, Boussinesq's, and Kelvin's

Chapter 4
The Flamant Problem

In 1892 [6], the French mechanist Alfred-Aimé Flamant (1839–1914) posed and solved the equilibrium problem of a linearly elastic, isotropic and homogeneous body occupying a half-space acted upon by a perpendicular *line load* of constant magnitude per unit length and infinitely long support (Fig. 4.1). In this chapter, we solve the Flamant Problem by a method different from his. Our method is semi-heuristic, in that its point of departure is a provisional representation for the solution fields that is suggested directly by physical intuition.

Although Flamant's line load mimics well the load exerted by a railroad track or a long foundation beam, in these and other applicative situations it is unrealistic to presume that the soil response be linearly elastic and isotropic. However, as we shall see, the Flamant's stress field does *not* depend on the material's constitution, and is therefore associable with whatever constitutive model is considered fit, in order to evaluate strains and displacements.

4.1 A Priori Representations of Displacement, Strain, and Stress

As a rule, finding the *data* $\mapsto$ *solution* mapping for a given problem is made easier when the data have implicit symmetries that the solution must reflect. Flamant's is a problem in 3-D classical elasticity where certain peculiar symmetries in the assignment of data do allow for an a priori parametric representation of the solution, which is then reduced to its explicit final form with ease.

Flamant's data are in many ways special: because of the geometry of the domain where the problem is formulated, the half-space

$$\mathcal{HS}^+ := \{x \mid (x - o) \cdot \boldsymbol{e}_1 \geq 0\};$$

because of the type of load, a force per unit length $\boldsymbol{f} = f\boldsymbol{e}_1$ which is uniformly distributed along the x_3 axis (Fig. 4.1); because of the material response, which is

P. Podio-Guidugli and A. Favata, *Elasticity for Geotechnicians*, Solid Mechanics and Its Applications 204, DOI: 10.1007/978-3-319-01258-2_4,

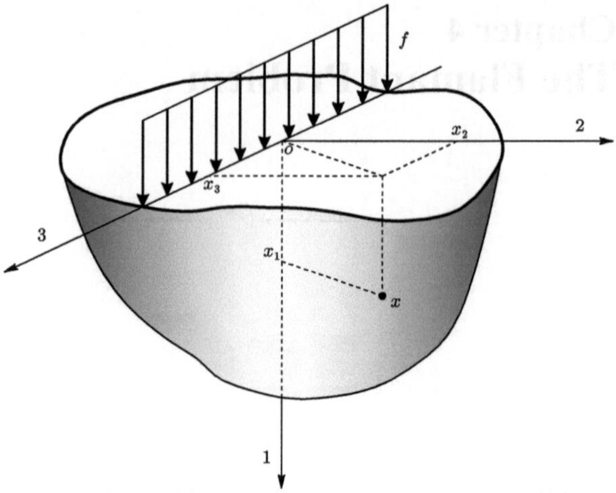

Fig. 4.1 The Flamant Problem

linearly elastic and isotropic. These peculiarities of the data induce our physical intuition to prefigure the qualitative features of both the displacement field and the stress field balancing the applied load: all we have to do is to give this prefigurement a convenient mathematical form.

As to the displacement field, we see that none of its components can depend on the coordinate x_3, because the load does not and the domain is unbounded; hence, the origin can be arbitrarily chosen on the x_3 axis:

$$\widetilde{\boldsymbol{u}}(\cdot,\cdot,x_3)=\widetilde{\boldsymbol{u}}(\cdot,\cdot,x_3+t)\quad\text{for each real number } t.$$

In particular, u_3, the component parallel to the load line, must be null all over the plane $x_3=0$, hence identically null all over $\mathcal{HS}^+$. Thus, the *solution displacement must be plane and independent of coordinate* x_3:

$$\boldsymbol{u}=\widetilde{u}_\alpha(x_1,x_2)\boldsymbol{e}_\alpha. \tag{4.1}$$

But this is not all: our physical intuition also suggests that we can limit ourselves to look for displacement fields being *mirror-symmetric with respect to plane* $x_2=0$:

$$\widetilde{u}_1(x_1,x_2)=\widetilde{u}_1(x_1,-x_2),\quad \widetilde{u}_2(x_1,x_2)=-\widetilde{u}_2(x_1,-x_2) \tag{4.2}$$

(therefore, in particular,

$$\widetilde{u}_2(x_1,0)=0$$

whatever the value of x_1, that is to say, at whatever distance from the surface of the half-space $\mathcal{HS}^+$).

Given that the displacement field (4.1) is plane, the Flamant strain field must be plane; consequently (recall the developments in Sect. 2.3.3), it must have the following form:

$$S_{\alpha\beta} = \widetilde{S}_{\alpha\beta}(x_1, x_2), \quad S_{3\alpha} \equiv 0,$$
$$S_{33} = \widetilde{S}_{33}(x_1, x_2) := \nu\big(\widetilde{S}_{11}(x_1, x_2) + \widetilde{S}_{22}(x_1, x_2)\big).$$

Both strain and stress fields must also inherit the mirror symmetry of the displacement field, detailed in (4.2).

Remark 4.1 The symmetry properties of the Flamant displacement field can be expressed also with the use of the *cylindrical coordinates* (ρ, ϑ, x_3) introduced in Sect. 3.2.2. Let us choose, for simplicity, $x_3 = 0$, and set:

$$\boldsymbol{u} = \widehat{\boldsymbol{u}}(\rho, \vartheta) := \widetilde{\boldsymbol{u}}(x_1, x_2),$$

for $\rho = (x_1^2 + x_2^2)^{1/2} \in (0, +\infty)$ and $\vartheta = \arccos(x_1/\rho) \in [-\pi/2, +\pi/2]$ the coordinates of a typical point of the half-plane

$$\mathcal{HP}^+ := \{x \mid (x - o) \cdot \boldsymbol{e}_3 = 0, \ \ (x - o) \cdot \boldsymbol{e}_1 \geq 0\}.$$

We see that we must have:

$$\begin{aligned} \widehat{\boldsymbol{u}}(\rho, \vartheta) \cdot \widehat{\boldsymbol{e}}(\vartheta) &= \widehat{\boldsymbol{u}}(\rho, -\vartheta) \cdot \widehat{\boldsymbol{e}}(-\vartheta), \\ \widehat{\boldsymbol{u}}(\rho, \vartheta) \cdot \widehat{\boldsymbol{e}}'(\vartheta) &= -\widehat{\boldsymbol{u}}(-\vartheta) \cdot \widehat{\boldsymbol{e}}'(-\vartheta), \end{aligned} \tag{4.3}$$

where the unit vectors $\boldsymbol{e}$ and $\boldsymbol{e}'$ are defined to be:

$$\begin{cases} \boldsymbol{e} = \widehat{\boldsymbol{e}}(\vartheta) := \cos\vartheta\, \boldsymbol{e}_1 + \sin\vartheta\, \boldsymbol{e}_2, \\ \boldsymbol{e}' = \widehat{\boldsymbol{e}}'(\vartheta) := -\sin\vartheta\, \boldsymbol{e}_1 + \cos\vartheta\, \boldsymbol{e}_2 \end{cases} \tag{4.4}$$

(Figure 4.2).

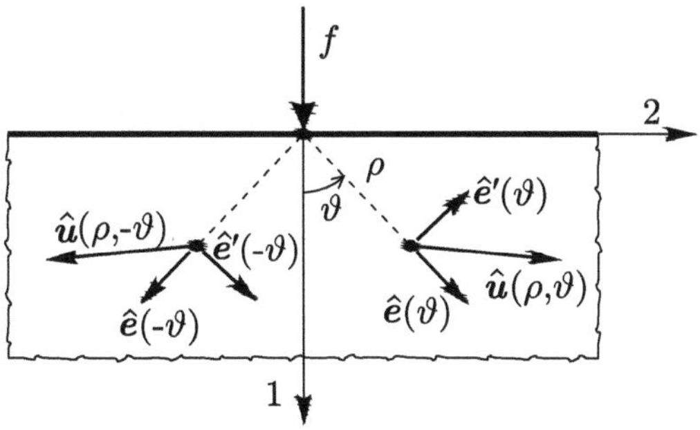

Fig. 4.2 Symmetries of Flamant's displacement field

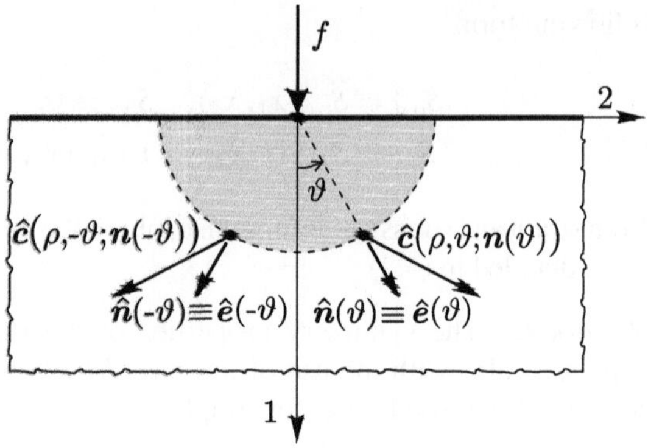

Fig. 4.3 Symmetries of Flamant's contact interactions

Remark 4.2 A representation in terms of cylindrical coordinates is better suited than a Cartesian representation to predict the form of the stress field. Consider the cylindric Euler cut, of radius ρ and axis x_3, whose cross section is visualized in Fig. 4.3. The part of half-space $\mathcal{HS}^+$ singled out by the cut, an infinite half-cylinder, must be in equilibrium under the combined action of the applied load and a distribution $\boldsymbol{c}$ of diffused contact actions exerted by the complementary part of $\mathcal{HS}^+$ at the common boundary, where

$$\widehat{\boldsymbol{n}}(\vartheta) = \widehat{\boldsymbol{e}}(\vartheta);$$

the stress field $\widehat{\boldsymbol{S}}$ associated with such distribution according to (2.29) must be such that:

$$\widehat{\boldsymbol{c}}(\rho, \vartheta; \widehat{\boldsymbol{n}}(\vartheta)) = \widehat{\boldsymbol{S}}(\rho, \vartheta)\widehat{\boldsymbol{e}}(\vartheta),$$

and our physical intuition suggests that the mapping $\widehat{\boldsymbol{c}}$ be given the same form restrictions that (4.3) details for $\widehat{\boldsymbol{u}}$.

4.2 Plane Stress Fields Balanced for Null Distance Forces

In this section, as a preliminary to our solution of the Flamant Problem, we construct a representation formula for plane stress fields such as those encountered when solving the plane versions not only of that problem but also of other similar problems, such as the plane Kelvin Problem studied in Sect. 6.2 and the Cerruti Problem studied in Chap. 8.

We know from Sect. 2.3.3 that a plane stress field $\widehat{\boldsymbol{S}}$ is termed balanced for null distance loads whenever it is divergenceless. We represent $\widehat{\boldsymbol{S}}$ as follows:

$$\begin{aligned}\boldsymbol{S} = \widehat{\boldsymbol{S}}(\rho, \vartheta) &= \widehat{\alpha}(\rho, \vartheta)\widehat{\boldsymbol{e}}(\vartheta) \otimes \widehat{\boldsymbol{e}}(\vartheta) + \widehat{\beta}(\rho, \vartheta)\widehat{\boldsymbol{e}}'(\vartheta) \otimes \widehat{\boldsymbol{e}}'(\vartheta)\\ &+ \widehat{\gamma}(\rho, \vartheta)(\widehat{\boldsymbol{e}}(\vartheta) \otimes \widehat{\boldsymbol{e}}'(\vartheta) + \widehat{\boldsymbol{e}}'(\vartheta) \otimes \widehat{\boldsymbol{e}}(\vartheta)).\end{aligned} \tag{4.5}$$

Then, in view of formula (3.18) for the divergence operator in polar coordinates, we find after some manipulations that, for $\widehat{\mathbf{S}}$ to be divergenceless in a plane region R, the scalar fields in the representation (4.5) must satisfy the following system of PDEs:

$$\begin{aligned}(\rho\alpha),_\rho &= \beta - \gamma,_\vartheta\,,\\ (\rho\gamma),_\rho &= -\beta,_\vartheta - \gamma,\end{aligned} \tag{4.6}$$

at each point of R. On adopting a heuristic attitude, we look for solutions of this system in the following separable form:

$$\widehat{\alpha}(\rho,\vartheta) = \alpha_0\,\rho^{-1}\widehat{a}(\vartheta), \quad \widehat{\gamma}(\rho,\vartheta) = \gamma_0\,\rho^{-1}\widehat{c}(\vartheta), \tag{4.7}$$

with the blanket assumption that functions $\widehat{a}$ and $\widehat{c}$ be as smooth as necessary for the developments to come to make sense. Under these hypotheses, system (4.6) reduces to:

$$\beta - \gamma,_\vartheta = 0, \quad \beta,_\vartheta + \gamma = 0,$$

implying that

$$\gamma,_{\vartheta\vartheta} + \gamma = 0,$$

or rather—equivalently, under the circumstances—that

$$\gamma_0\big(\widehat{c}''(\vartheta) + \widehat{c}(\vartheta)\big) = 0.$$

Hence,

$$\widehat{c}(\vartheta) = c_1\cos\vartheta + c_2\sin\vartheta \quad \text{and} \quad \widehat{\beta}(\rho,\vartheta) = \gamma_0\,\rho^{-1}(-c_1\sin\vartheta + c_2\cos\vartheta).$$

Summing up, there is a family of divergenceless plane stress fields having the following separable representation of type (4.5):

$$\begin{aligned}\rho\widehat{\mathbf{S}}(\rho,\vartheta) &= \alpha_0\,\widehat{a}(\vartheta)\widehat{\mathbf{e}}(\vartheta)\otimes\widehat{\mathbf{e}}(\vartheta)\\ &\quad + \gamma_0\Big(\widehat{c}'(\vartheta)\widehat{\mathbf{e}}'(\vartheta)\otimes\widehat{\mathbf{e}}'(\vartheta) + \widehat{c}(\vartheta)\big(\widehat{\mathbf{e}}(\vartheta)\otimes\widehat{\mathbf{e}}'(\vartheta) + \widehat{\mathbf{e}}'(\vartheta)\otimes\widehat{\mathbf{e}}(\vartheta)\big)\Big),\end{aligned} \tag{4.8}$$

a family parameterized by the three constants α_0, $\gamma_0 c_1$, $\gamma_0 c_2$ and by the scalar function $\widehat{a}$.

Remark 4.3 In the next section we shall show that the stress field solving the plane Flamant Problem obtains from (4.8) on choosing

$$\gamma_0 = 0, \tag{4.9}$$

whence

$$\widehat{\mathbf{S}}(\rho,\vartheta) = \alpha_0\,\rho^{-1}\widehat{a}(\vartheta)\widehat{\mathbf{e}}(\vartheta)\otimes\widehat{\mathbf{e}}(\vartheta). \tag{4.10}$$

The reason why γ_0 is taken null is that otherwise it would be impossible to satisfy the null-traction condition at all points but the origin of the straight line that bounds the half-plane $\mathcal{HP}^+$. Indeed, with reference to Fig. 4.3, for $\rho > 0$ and by continuity up to $\vartheta = \pm\pi/2$, (4.8) yields the stress field:

$$\begin{aligned}\rho\widehat{\boldsymbol{S}}(\rho, \pm\pi/2) &= \alpha_0\,\widehat{a}(\pm\pi/2)\,\boldsymbol{e}_2 \otimes \boldsymbol{e}_2\\ &\quad + \gamma_0\big(\widehat{c}'(\pm\pi/2)\,\boldsymbol{e}_1 \otimes \boldsymbol{e}_1 - \widehat{c}(\pm\pi/2)(\boldsymbol{e}_2 \otimes \boldsymbol{e}_1 + \boldsymbol{e}_1 \otimes \boldsymbol{e}_2)\big)\\ &= \alpha_0\,\widehat{a}(\pm\pi/2)\,\boldsymbol{e}_2 \otimes \boldsymbol{e}_2 \mp \gamma_0\big(c_1\,\boldsymbol{e}_1 \otimes \boldsymbol{e}_1 + c_2(\boldsymbol{e}_2 \otimes \boldsymbol{e}_1 + \boldsymbol{e}_1 \otimes \boldsymbol{e}_2)\big),\end{aligned}$$

and hence, the traction vector is:

$$\boldsymbol{c}(\rho, \pm\pi/2; \boldsymbol{e}_1) = -\widehat{\boldsymbol{S}}(\rho, \pm\pi/2)\boldsymbol{e}_1 = \pm\gamma_0\rho\big(c_1\,\boldsymbol{e}_1 + c_2\,\boldsymbol{e}_2\big).$$

Hence,

$$\boldsymbol{c}(\rho, \pm\pi/2; \boldsymbol{e}_1) \equiv \boldsymbol{0} \;\; (\rho > 0) \quad \Leftrightarrow \quad \gamma_0 = 0.$$

4.3 The 2-D Boussinesq–Flamant Problem

In 1878 [2], Joseph Valentin Boussinesq (1842–1929)—he too a student of Saint-Venant's, just as Flamant—had considered the case of a point-concentrated load perpendicular to a half-space, a problem he was to return to repeatedly later in [3–5]. We shall study the Boussinesq Problem in the next chapter. A glance to Figs. 5.1 and 4.1 is enough to conclude that the 2-D versions of that problem and Flamant's are no different. For this reason, we name after both Boussinesq and Flamant the plane equilibrium problem we study in this section.

4.3.1 Divergenceless Plane Stress Fields

We denote by $\boldsymbol{r} := x - o$ the position vector of point $x \in \mathcal{HP}^+$ with respect to o; moreover, we set $\rho := |\boldsymbol{r}|$, $\boldsymbol{e} := \rho^{-1}\boldsymbol{r}$, $\vartheta := \arcsin(\boldsymbol{e}_1 \times \boldsymbol{e} \cdot \boldsymbol{e}_3)$, whence, in particular, $(4.4)_1$ follows. Our first goal is to find out what stress fields balance the applied force $\boldsymbol{f} = f\boldsymbol{e}_1$.

One may ask why the separate-variable representations we chose in the preceding section for the parameter functions $\widehat{\alpha}$ and $\widehat{\gamma}$ featured a first-order singularity in ρ at the origin. With a view toward motivating this choice, we propose to start from the following *Ansatz*:

$$\boldsymbol{S} = \widehat{\sigma}(\rho, \vartheta)\widehat{\boldsymbol{e}}(\vartheta) \otimes \widehat{\boldsymbol{e}}(\vartheta), \tag{4.11}$$

with $\widehat{\sigma}(\rho, \cdot)$ an *even* function of ϑ. As a matter of fact, at the periphery of the half-disk $\mathcal{HD}_\rho$ of radius ρ depicted in Fig. 4.4, where the exterior normal coincides with the

Fig. 4.4 The half-disk $\mathcal{HD}_\rho$

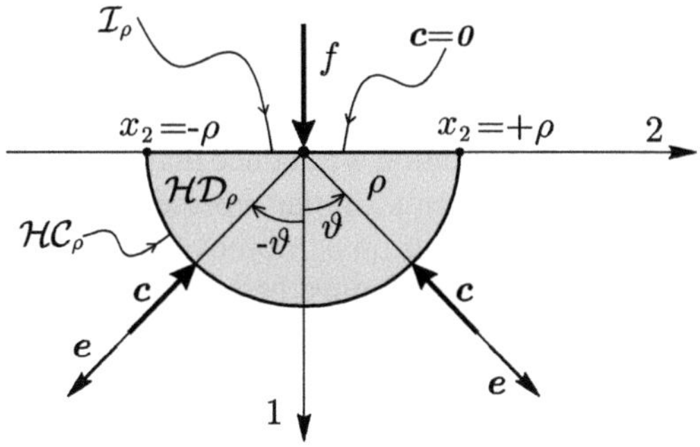

unit vector $\boldsymbol{e}$, our physical intuition suggests that the stress field (4.11) induces the contact interaction

$$\widehat{\boldsymbol{c}}(\rho, \vartheta; \boldsymbol{n}) = \widehat{\boldsymbol{S}}(\rho, \vartheta)\boldsymbol{n} = (\boldsymbol{n} \cdot \widehat{\boldsymbol{e}}(\vartheta))\widehat{\sigma}(\rho, \vartheta)\widehat{\boldsymbol{e}}(\vartheta), \quad \widehat{\sigma}(\rho, \vartheta) = \widehat{\sigma}(\rho, -\vartheta), \quad \forall \rho > 0;$$

in particular, over the half-circumference

$$\mathcal{HC}_\rho := \{x \mid x - o = \rho\widehat{\boldsymbol{e}}(\vartheta),\ \vartheta \in (-\pi/2, +\pi/2)\},$$

the vector field $\widehat{\boldsymbol{c}}$ is radial:

$$\widehat{\boldsymbol{c}}(\rho, \vartheta; \boldsymbol{n}) = \widehat{\sigma}(\rho, \vartheta)\widehat{\boldsymbol{e}}(\vartheta),$$

while it is null over the segment

$$\mathcal{I}_\rho := \{x \mid x - o = \sigma\boldsymbol{e}_2,\ \sigma \in (+\rho, 0) \cup (0, -\rho)\}.$$

It follows from (4.11) that

$$\operatorname{div} \boldsymbol{S} = (\sigma,_\rho + \rho^{-1}\sigma)\boldsymbol{e},$$

whence

$$\sigma,_\rho + \rho^{-1}\sigma = 0; \Leftrightarrow; \widehat{\sigma}(\rho, \vartheta) = \rho^{-1}\widehat{\tau}(\vartheta), \ \text{with}\ \widehat{\tau}(\vartheta) = \widehat{\tau}(-\vartheta).$$

Thus, the *Ansatz* (4.11) is further specified by giving it the form (4.10), with $\widehat{a}$ even:

$$\boldsymbol{S} = \alpha_0\rho^{-1}\widehat{a}(\vartheta)\widehat{\boldsymbol{e}}(\vartheta) \otimes \widehat{\boldsymbol{e}}(\vartheta), \quad \widehat{a}(\vartheta) = \widehat{a}(-\vartheta). \tag{4.12}$$

Remark 4.4 If function $\widehat{\alpha}$ in (4.8) is chosen as in $(4.7)_1$, then system (4.6) becomes:

$$-\beta + \gamma,_\vartheta = 0,$$
$$\gamma,_\rho + \rho^{-1}(\beta,_\vartheta + 2\gamma) = 0.$$

From the first equation we deduce that, if a separate-variable representation is sought for function $\widehat{\gamma}$, then a similar representation must hold also for function $\widehat{\beta}$, with an identical dependence on ρ. Moreover, by combining the two equations in the system, we see that function $\widehat{\gamma}$ must be such that

$$(\rho\gamma),_\rho + \gamma,_{\vartheta\vartheta} + \gamma = 0.$$

From this last relation we infer that $\widehat{\gamma}$, to have a separate-variable representation, must be of the form $(4.7)_2$. Therefore, due to $(4.6)_1$, $\widehat{\beta}$ must be of the form $(4.7)_2$ as well.

4.3.2 Compatible and Divergenceless Plane Stress Fields

We now wish to find out for which choice of the parameter function $\widehat{a}$ a divergenceless plane stress field of type (4.12) is *compatible*, that is, induces a plane strain field in the half-plane $\mathcal{HP}^+$ when that half-plane is comprised of a linearly elastic isotropic material. To this end—we know from Sect. 2.4.2—it is sufficient that the field $\widehat{\boldsymbol{S}}$ satisfies Eq. (2.69):

$$\Delta(\operatorname{tr} \boldsymbol{S}) = 0,$$

that is, in the present instance,

$$\operatorname{tr} \boldsymbol{S} = \alpha_0 \rho^{-1} \widehat{a}(\vartheta).$$

Hence, on recalling $(3.19)_3$, function $\widehat{a}$ must such that

$$\Delta(\rho^{-1}\widehat{a}(\vartheta)) = \rho^{-2}(\widehat{a}''(\vartheta) + \widehat{a}(\vartheta)) = 0.$$

To conclude without omitting to satisfy also $(4.12)_2$, we must choose

$$\widehat{a}(\vartheta) = \cos\vartheta,$$

that is, the solution (unique to within an inessential multiplicative constant) to the following problem:

$$a''(\vartheta) + \widehat{a}(\vartheta) = 0, \quad \widehat{a}(\vartheta) = \widehat{a}(-\vartheta), \quad \vartheta \in (-\pi/2, +\pi/2).$$

4.3.3 Compatible and Divergenceless Plane Stress Fields that Balance the Applied Force

To complete our construction of the balanced plane stress field that solves the Boussinesq-Flamant Problem we have to find the value of the constant α_0. This we do by an imposition of part-wise equilibrium.

For any fixed $\rho > 0$, we consider the half-disk $\mathcal{HD}_\rho$ depicted in Fig. 4.4: on its boundary, it follows from (4.12) that the stress vector field is null at points of $\mathcal{I}_\rho$ and is

$$\boldsymbol{S}\boldsymbol{e} = \alpha_0 \rho^{-1} \cos\vartheta\, \boldsymbol{e} \tag{4.13}$$

at points of $\mathcal{HC}_\rho$. Therefore, $\mathcal{HD}_\rho$ is in equilibrium if

$$\int_{\mathcal{HC}_\rho} \boldsymbol{S}\boldsymbol{e} + \boldsymbol{f} = \boldsymbol{0}.$$

Now,

$$\begin{aligned}\int_{\mathcal{HC}_\rho} \boldsymbol{S}\boldsymbol{e} &= \alpha_0 \int_{-\pi/2}^{+\pi/2} \Big(\rho^{-1}\cos\vartheta(\cos\vartheta\boldsymbol{e}_1 + \sin\vartheta\boldsymbol{e}_2)\Big)\rho\, d\vartheta \\ &= \alpha_0\Big(\int_{-\pi/2}^{+\pi/2} \cos^2\vartheta d\vartheta\Big)\boldsymbol{e}_1 = \alpha_0\frac{\pi}{2}\boldsymbol{e}_1 \quad \text{and} \quad \boldsymbol{f} = f\boldsymbol{e}_1;\end{aligned}$$

hence,

$$\alpha_0 = -\frac{2f}{\pi}.$$

In conclusion, the *Boussinesq-Flamant stress field* in $\mathcal{HP}^+$ is:

$$\boldsymbol{S}^{BF} = \widehat{\boldsymbol{S}}^{BF}(\rho,\vartheta) := -\frac{2f}{\pi}\rho^{-1}\cos\vartheta\,\widehat{\boldsymbol{e}}(\vartheta)\otimes\widehat{\boldsymbol{e}}(\vartheta), \tag{4.14}$$

for all $(\rho,\vartheta) \in (0,+\infty)\times[-\pi/2,+\pi/2]$.[1]

Remark 4.5 The appropriate restriction of the field $\widehat{\boldsymbol{S}}^{BF}$ solves the equilibrium problem for a *wedge*

$$\mathcal{W}_{\vartheta_0} := \{x \mid x - o = \rho\widehat{\boldsymbol{e}}(\vartheta),\ \widehat{\boldsymbol{e}}(\vartheta)\cdot\boldsymbol{e}_3 = 0,\ |\widehat{\boldsymbol{e}}(\vartheta)\cdot\boldsymbol{e}_1| \le |\cos\vartheta_0|\},$$

whatever the vertex angle $2\vartheta_0 \in (0, 2\pi)$ (Fig. 4.5); the stress field in the wedge turns out to be:

[1] See Sect. A.3.1 for an exposition of the classical Airy method to construct balanced and compatible plane stress fields.

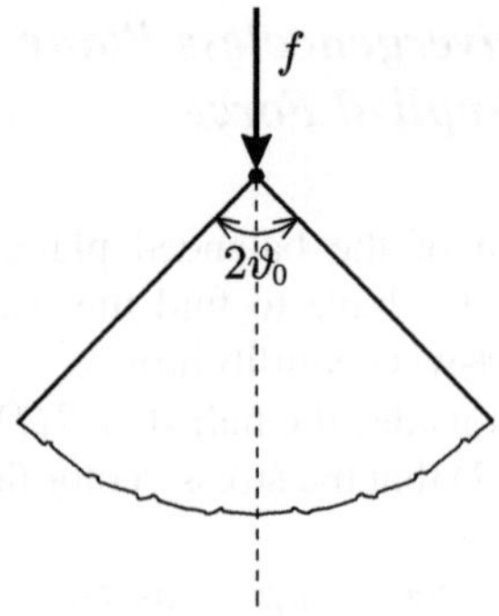

Fig. 4.5 A symmetrically-loaded wedge of opening $2\vartheta_0$

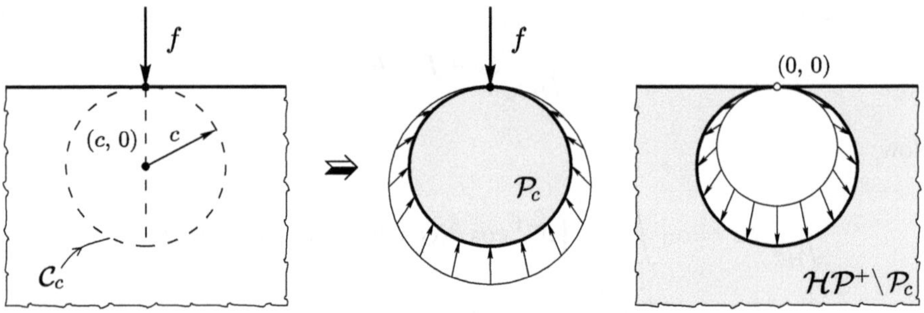

Fig. 4.6 All along the locus of constant stress-magnitude, contact interactions are exclusively diffused (this figure is taken from [8])

$$\mathbf{S}_{\mathcal{W}} = -\frac{f}{\vartheta_0\,\rho}\cos\vartheta\,\widehat{\mathbf{e}}(\vartheta)\otimes\widehat{\mathbf{e}}(\vartheta), \quad (\rho,\vartheta)\in(0,+\infty)\times[-\vartheta_0,+\vartheta_0].$$

Remark 4.6 The magnitude of $\mathbf{S}^{BF}$ is constant at those points of $\mathcal{HP}^+$ where

$$\rho^{-1}\cos\vartheta = (2c)^{-1} = \text{a positive constant;}$$

it is not difficult to see that those points lie on the circumference $\mathcal{C}_c$ of a circle of center $(c, 0)$ and radius c (Fig. 4.6).[2] At a point of $\mathcal{C}_c$, the outer unit normal is

$$\mathbf{n} = \cos\vartheta\,\mathbf{e} + \sin\vartheta\,\mathbf{e}'$$

(see Fig. 4.7). Hence, the stress vector turns out to be

---

[2] This circumference has the Cartesian equation

$$c^2 = (x_1 - c)^2 + x_2^2 = (\rho\cos\vartheta - c)^2 + (\rho\sin\vartheta)^2 = \rho^2(1 - (2c)\rho^{-1}\cos\vartheta + c^2/\rho^2).$$

In the geotechnical literature, this locus is called the *pressure bulb*.

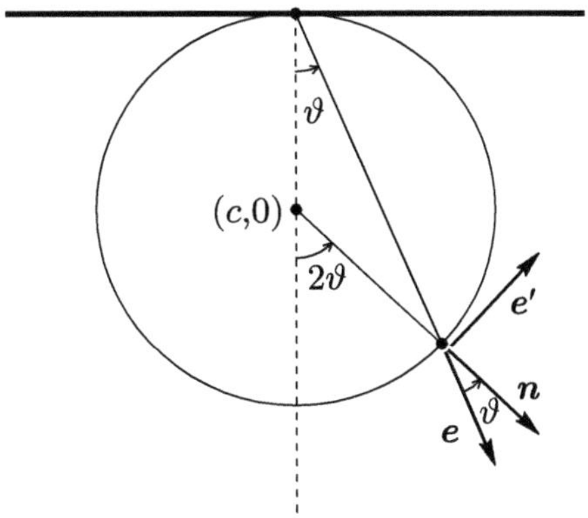

Fig. 4.7 The outer normal at a typical point of circimference $\mathcal{C}_c$

$$S^{BF}n = -\frac{f}{\pi c}\cos\vartheta e. \tag{4.15}$$

We think of the part of $\mathcal{HP}^+$ bounded by $\mathcal{C}_c$ as being balanced under the combined action of the diffused force (4.15) and the concentrated load $\boldsymbol{f}$; consistently, we think of the complementary part as subject to diffused contact interactions only (Fig. 4.6).

4.4 Digression

(This section is taken almost *verbatim* from [8].)

4.4.1 A Weak Formulation of the Boussinesq–Flamant Problem

The stress field $\widehat{S}^{BF}$ satisfies point-wise the force balance equation:

$$\operatorname{div} S^{BF} = 0 \tag{4.16}$$

in all of $\mathcal{HP}^+ \setminus \{o\}$; we now consider in what sense it satisfies the accompanying traction boundary conditions, for the half-space and for its parts.

We propose to interpret the force $\boldsymbol{f}$ concentrated at o as a *Dirac traction* $\boldsymbol{t}_\rho$ applied over the segment $\mathcal{I}_\rho$:

$$\boldsymbol{f} = \int_{\mathcal{I}_\rho} \boldsymbol{t}_\rho, \quad \boldsymbol{t}_\rho(x) := f\delta(x-o)\boldsymbol{e}_1, \; x \in \mathcal{I}_\rho;$$

with this interpretation, the relation

$$\int_{\mathcal{HC}_\rho} \boldsymbol{S}^{BF}\boldsymbol{e} + \int_{\mathcal{I}_\rho} \boldsymbol{t}_\rho = \boldsymbol{0}$$

expresses the *balance of contact actions*, internal and external, applied to part $\mathcal{HD}_\rho$, whatever $\rho > 0$. Moreover, it follows from a formally stated divergence theorem and (4.16) that

$$\int_{\mathcal{HD}_\rho} \operatorname{div} \boldsymbol{S}^{BF} = \int_{\mathcal{HC}_\rho} \boldsymbol{S}^{BF}\boldsymbol{e} + \int_{\mathcal{I}_\rho} \boldsymbol{S}^{BF}\boldsymbol{e}_1 = \int_{\mathcal{I}_\rho} (-\boldsymbol{t}_\rho + \boldsymbol{S}^{BF}\boldsymbol{e}_1) = \boldsymbol{0}. \quad (4.17)$$

We regard the last relation as an appropriate weak version of the *traction boundary condition* prevailing over the segment $\mathcal{I}_\rho$ [7].

The last equality in (4.17) shows that, over the straight line bounding $\mathcal{HP}^+$, the surface traction $\boldsymbol{S}^{BF}\boldsymbol{n} = -\boldsymbol{S}^{BF}\boldsymbol{e}_1$ is in the parlance of integration theory a *measure*, concentrated at the point o where the external *contact force* $\boldsymbol{f}$ is applied [11]. Suppose now that the field $\widetilde{\boldsymbol{S}}^{BF}$ is continuously extended to null to the upper half-plane $\mathcal{H}^-$, and let $\widetilde{\boldsymbol{S}}^{BF}$ denote such extended field over the plane $\mathcal{P} = \mathcal{HP}^+ \cup \mathcal{HP}^-$. Interestingly, the field $\widetilde{\boldsymbol{S}}^{BF}$ has *divergence measure* in $\mathcal{P}$, equal to $-\delta(x-o)\boldsymbol{f},\ x \in \mathcal{P}$.[3] To interpret the latter result along the same lines we have interpreted the former, i.e., as a consequence of a force balance, we may consider a disk-shaped part $\mathcal{D}_\rho$ of $\mathcal{P}$, of center o and radius ρ, and imagine it as subject to an external *distance force* $\boldsymbol{f}$ applied at o, balanced by diffused tractions being identically null over $\partial\mathcal{D}_\rho \cap \mathcal{HP}^-$ and equal to $\boldsymbol{S}^{BF}\boldsymbol{e}$ over $\partial\mathcal{D}_\rho \cap \mathcal{HP}^+$. Continuum mechanics provides us with a unifying format for balance statements that allows for a further, precise interpretation of these two analytical findings: A pair $\big((\boldsymbol{c}, \boldsymbol{d}), \boldsymbol{S}\big)$, formed by contact and distance force fields $\boldsymbol{c}$ and $\boldsymbol{d}$ and a stress field $\boldsymbol{S}$ over a region Ω with boundary $\partial\Omega$, is *weakly balanced* whenever the stress working equals the distance working plus the contact working, i.e., whenever

$$\int_\Omega \boldsymbol{S} \cdot \nabla \boldsymbol{v} = \int_\Omega \boldsymbol{d} \cdot \boldsymbol{v} + \int_{\partial\Omega} \boldsymbol{c} \cdot \boldsymbol{v}, \quad \text{for all smooth test fields } \boldsymbol{v};$$

moreover, in view of a standard differential identity,

$$\int_\Omega \boldsymbol{S} \cdot \nabla \boldsymbol{v} = \int_\Omega (-\operatorname{div} \boldsymbol{S}) \cdot \boldsymbol{v} + \int_{\partial\Omega} (\boldsymbol{S}\boldsymbol{n}) \cdot \boldsymbol{v},$$

so that we can regard $-\operatorname{div} \boldsymbol{S}$ as the distance force, and $\boldsymbol{S}\boldsymbol{n}$ as the contact force, associated to a given stress field $\boldsymbol{S}$. If we apply this balance format to the parts $\mathcal{HD}_\rho$ of $\mathcal{HP}^+$ and $\mathcal{D}_\rho$ of $\mathcal{P}$, we find that

$$\int_{\mathcal{HD}_\rho} \boldsymbol{S}^{BF} \cdot \nabla \boldsymbol{v} - \int_{\mathcal{HC}_\rho} \boldsymbol{S}^{BF}\boldsymbol{e} \cdot \boldsymbol{v} = \boldsymbol{f} \cdot \boldsymbol{v}(o) = \int_{\mathcal{D}_\rho} \widetilde{\boldsymbol{S}}^{BF} \cdot \nabla \boldsymbol{v} - \int_{\mathcal{HC}_\rho} \boldsymbol{S}^{BF}\boldsymbol{e} \cdot \boldsymbol{v};$$

[3] A. Musesti, private communication, June 2004.

in plain words, that the working of the external force applied at o equals the difference between the stress working and the contact working over $\mathcal{HC}_\rho$. In the case of $\mathcal{P}_\rho$, we may write

$$\boldsymbol{f}\cdot\boldsymbol{v}(o)=\int_{\mathcal{I}_\rho}(\mathbf{S}^{BF}\boldsymbol{n})\cdot\boldsymbol{v}=\text{the contact working over}\,\mathcal{I}_\rho;$$

in the case of $\mathcal{D}_\rho$,

$$\boldsymbol{f}\cdot\boldsymbol{v}(o)=\int_{\mathcal{D}_\rho}(-\mathrm{div}\,\widetilde{\mathbf{S}}^{BF})\cdot\boldsymbol{v}=\text{the distance working over}\,\mathcal{D}_\rho.$$

4.4.2 Concentrated Contact Interactions

That concentrated contact interactions are in order between adjacent parts of the Boussinesq-Flamant half-plane is easily demonstrated by the use of free-body diagrams.

Consider the quarter-disk $\mathcal{QD}_\rho$ sketched in Fig. 4.8. When part $\mathcal{QD}_\rho$ is ideally cut away from the rest of $\mathcal{HP}^+$, then it must be in equilibrium under the action of: (i) the concentrated force $1/2\,\boldsymbol{f}\boldsymbol{e}_1$; (ii) the diffused contact force $\mathbf{S}^{BF}\boldsymbol{e}$ exerted by the right adjacent part $\mathcal{Q}^{(r)}$ (the same as the internal contact interaction between the two parts before the cut!), which, in view of (4.13), is equipollent to the concentrated force

$$\int_{\frac{1}{2}\mathcal{HC}_\rho}\mathbf{S}^{BF}\boldsymbol{e}=-\frac{1}{2}\boldsymbol{f}\boldsymbol{e}_1-\frac{1}{\pi}\boldsymbol{f}\boldsymbol{e}_2,$$

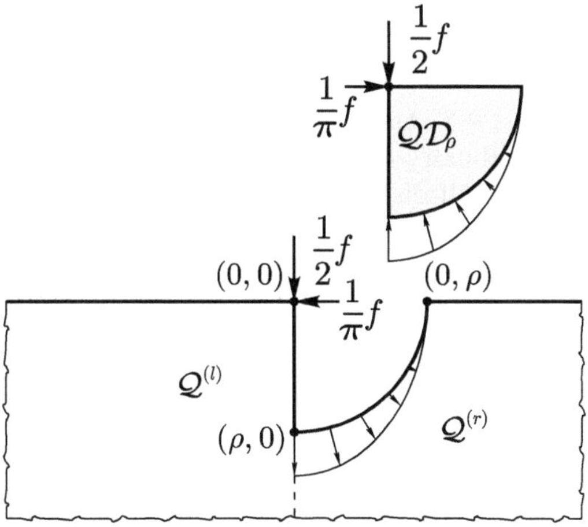

Fig. 4.8 For a quarter-disk, a concentrated contact interaction is needed to balance the applied load (this figure is taken from [8])

applied at o; (iii) the contact action exerted by the left adjacent part $\mathcal{Q}^{(l)}$. Now, the diffused contact force $-\boldsymbol{S}^{BF}\boldsymbol{e}_2$ exerted by $\mathcal{Q}^{(l)}$ on $\mathcal{QD}_\rho$ is everywhere null along their common boundary:

$$\widehat{\boldsymbol{S}}^{BF}(\sigma, 0)\boldsymbol{e}_2 \equiv 0 \quad \text{for} \quad \sigma \in [0, \rho]\,,$$

just as their internal contact interaction is before the cut. Then, to guarantee the free-body equilibrium of $\mathcal{QD}_\rho$, we are driven to admit that the cut operation brings into evidence an internal *concentrated contact interaction*

$$\widehat{\boldsymbol{f}}(\mathcal{QD}_\rho, \mathcal{Q}^{(l)}) = -\widehat{\boldsymbol{f}}(\mathcal{Q}^{(l)}, \mathcal{QD}_\rho) = \frac{1}{\pi} f\boldsymbol{e}_2$$

at point o.[4] For another example of concentrated contact interactions, see Remark 8.1.

Remark 4.7 This result is neither dependent on the parameter ρ nor on whichever curve from point $(\rho, 0)$ to point $(0, \rho)$ we pick to bound a body part alternative to $\mathcal{QD}_\rho$. Indeed, if $\boldsymbol{r} = r(\vartheta)\boldsymbol{e}(\vartheta)$ is the position vector of a typical point x on such a curve, denoted by $\mathcal{C}$ in Fig. 4.9, then the vectors $\boldsymbol{r}' = r'\boldsymbol{e} + r\boldsymbol{e}'$ and $\boldsymbol{n} = |\boldsymbol{r}'|^{-1}\boldsymbol{r}' \times \boldsymbol{e}_3$ are, respectively, tangent and normal to $\mathcal{C}$; in particular, then, $\boldsymbol{e} \cdot \boldsymbol{n} = |\boldsymbol{r}'|^{-1}r$. The total contact action exerted along any chosen portion of the curve $\mathcal{C}$ is then the same as the contact action exerted on the corresponding portion of $\frac{1}{2}\mathcal{HC}_\rho$:

$$\int_{[\mathcal{C}]_{\vartheta_0}^{\vartheta_1}} \boldsymbol{S}^{BF}\boldsymbol{n} = \int_{\vartheta_0}^{\vartheta_1} (\boldsymbol{e}\cdot\boldsymbol{n})\boldsymbol{S}^{BF}\boldsymbol{e}|\boldsymbol{r}'|d\vartheta = \int_{\vartheta_0}^{\vartheta_1} \boldsymbol{S}^{BF}\boldsymbol{e}\,r\,d\vartheta = \int_{[\frac{1}{2}\mathcal{HC}_\rho]_{\vartheta_0}^{\vartheta_1}} \boldsymbol{S}^{BF}\boldsymbol{e}\,.$$

Thus, the concentrated contact force arising at point o is a *local effect*, in the sense that it is a manifestation (in this case, the only manifestation) of the interaction between any two adjacent body parts sharing the segment $\{x \mid x - o = \sigma\boldsymbol{e}_1,\ \sigma \in [0, \rho]\}$ as a common boundary, whatever $\rho > 0$. Note also that this effect concentrates at a point that belongs to the *topological boundary* of part $\mathcal{P}_\rho$, but not to its *reduced boundary*.[5]

Remark 4.8 Assuming that the vertical external force on part $\mathcal{QD}_\rho$ be $1/2\, f$ may seem arbitrary and ponderous, but in fact it is not. To see this, imagine to ideally cut the half-disk of Fig. 4.4 into two identical quarter-disks, with a view toward sketching a free-body diagram for each of the latter: symmetry then requires that the concentrated external force is split equal.

Remark 4.9 That the concentrated interaction forces at the vertex of the right angle, both in part $\mathcal{HD}_\rho$ and in its complement, should be those shown in Fig. 4.8 can be

---

[4] Here $\widehat{\boldsymbol{f}}(\mathcal{A}, \mathcal{B})$ denotes the total contact force exerted by part $\mathcal{B}$ over part $\mathcal{A}$ along their common boundary.

[5] Roughly speaking, the reduced boundary of a set—a measure-theoretic notion carefully introduced, e.g., in [1], p. 154—is the subset of all points of the topological boundary where a (inner) normal is well defined.

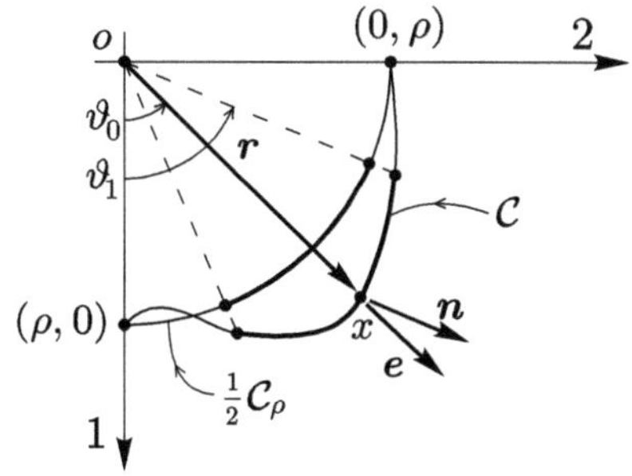

Fig. 4.9 The concentrated contact interaction is the same whatever the curve $\mathcal{C}$ joining $(\rho, 0)$ to $(0, \rho)$ (this figure is taken from [8])

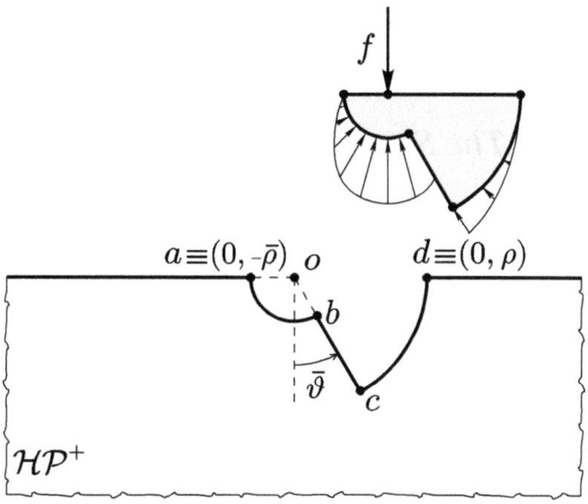

Fig. 4.10 Free-body diagram for a part of the Boussinesq-Flamant half-plane in the form of a "nosed piece of pie" (this figure is taken from [8])

seen also by a limit argument suggested by R. Fosdick.[6] With reference to Fig. 4.10, we let

$$\boldsymbol{g}(\bar{\rho}, \bar{\vartheta}) = -\int_{[\mathcal{C}_\rho]_{-\frac{\pi}{2}}^{\bar{\vartheta}}} \mathbf{S}^{BF}\boldsymbol{e}$$

denote the total force equipollent with respect to point o to the diffused interaction force exerted by the "nosed" part on the right over its complement along the common boundary curve through points a, b, and c; and we compute

$$\boldsymbol{g}(\bar{\rho}, \bar{\vartheta}) = \frac{1}{2} f\left(1 + \frac{2}{\pi}(\bar{\vartheta} + \sin\bar{\vartheta}\cos\bar{\vartheta})\right)\boldsymbol{e}_1 - \frac{1}{\pi} f\left(1 - \sin^2\bar{\vartheta}\right)\boldsymbol{e}_2.$$

Not surprisingly, in the light of Remark 4.7, this force is independent of the parameter $\bar{\rho}$, and reduces to the expected vector for $\bar{\vartheta} = 0$.

Remark 4.10 Once again, it is appropriate here to attract the reader's attention on two papers by F. Schuricht, were the construction of an interaction theory general

[6] Private communication, August 2004.

enough to allow for concentrated contact interactions between adjacent body parts has been undertaken [9, 10].

4.5 The Flamant Elastic State

As repeatedly anticipated, our approach to solving the Flamant Problem is different from the standard one, where the strain and stress fields are computed after the displacement field has been found. Instead, we proceed in reverse order, and derive the stress field first, then the deformation field, lastly the displacement field.

4.5.1 The Stress Field

To guarantee that the 3-D strain and displacement fields both come out plane, we add to the plane stress field $\boldsymbol{S}^{BF}$ given by (4.14)—that is, to the stress field that solves the 2-D Boussinesq-Flamant Problem—a stress field having only one nonnull component, namely,

$$S_{33} = \nu \,\mathrm{tr}\, \boldsymbol{S}^{BF}$$

(cf. 2.70). The result is the 3-D stress field:

$$\boldsymbol{S} = \widetilde{\boldsymbol{S}}^F(\rho, \vartheta, x_3) = -\frac{2f}{\pi}\rho^{-1}\cos\vartheta\Big(\widehat{\boldsymbol{e}}(\vartheta) \otimes \widehat{\boldsymbol{e}}(\vartheta) + \nu \boldsymbol{e}_3 \otimes \boldsymbol{e}_3\Big). \tag{4.18}$$

We see at once that this field is balanced for everywhere null distance forces:

$$\mathrm{div}\, \boldsymbol{S} = 0;$$

that, over all the $x_1 = 0$ plane minus the line where the load is applied,

$$\boldsymbol{S}\boldsymbol{n} = -\boldsymbol{S}\boldsymbol{e}_1 \equiv 0;$$

and that

$$\lim_{\rho\to+\infty} \widetilde{\boldsymbol{S}}^F(\rho, \vartheta, x_3) = 0, \quad \text{whatever } (\vartheta, x_3).$$

4.5.2 The Strain Field

There are two ways to obtain the plane strain field that solves the Flamant Problem: on inserting the plane stress field $\boldsymbol{S}^{BF}$ into the inverse constitutive Eq. (2.57), one finds that

$$\boldsymbol{E} = \frac{1}{E_0}\Big((1+\nu_0)\boldsymbol{S}^{BF} - \nu_0(\operatorname{tr}\boldsymbol{S}^{BF})\boldsymbol{I}_{(2)}\Big); \tag{4.19}$$

alternatively, one can move from the 3-D stress field (4.18) and make use of (2.45). Combination of (4.19) and (4.14) gives:

$$\boldsymbol{E} = -\frac{2f}{\pi E_0}\rho^{-1}\cos\vartheta\Big((1+\nu_0)\boldsymbol{e}\otimes\boldsymbol{e} - \nu_0\boldsymbol{I}_{(2)}\Big),$$

where

$$\boldsymbol{I}_{(2)} = \boldsymbol{e}\otimes\boldsymbol{e} + \boldsymbol{e}'\otimes\boldsymbol{e}';$$

hence,

$$\boldsymbol{E} = \widetilde{\boldsymbol{E}}^F(\rho, \vartheta, x_3) = -\frac{2f}{\pi E_0}\rho^{-1}\cos\vartheta\Big(\boldsymbol{e}\otimes\boldsymbol{e} - \nu_0\boldsymbol{e}'\otimes\boldsymbol{e}'\Big).$$

We see that the strain field consists of a *radial contraction*

$$E_{\rho\rho} = \boldsymbol{E}\boldsymbol{e}\cdot\boldsymbol{e} = -\frac{2f}{\pi E_0}\rho^{-1}\cos\vartheta, \tag{4.20}$$

accompanied by an additional deformation:

$$E_{\vartheta\vartheta} = \boldsymbol{E}\boldsymbol{e}'\cdot\boldsymbol{e}' = \nu_0\frac{2f}{\pi E_0}\rho^{-1}\cos\vartheta, \tag{4.21}$$

that turns out to be a *circumferential dilatation* if the Poisson's modulus is positive.[7] Both radial contraction and circumferential dilatation decay with distance from the point where the load is applied; as is obvious in view of the problem's symmetries, there is no change in angle between radial and circumferential fibers:

$$E_{\rho\vartheta} = E_{\vartheta\rho} = \boldsymbol{E}\boldsymbol{e}\cdot\boldsymbol{e}' = 0.$$

Finally, we observe that the *area change* in the plane perpendicular to $\boldsymbol{e}_3$ is:

$$E_{\rho\rho} + E_{\vartheta\vartheta} = -\frac{2f}{\pi E_0}\rho^{-1}\cos\vartheta(1-\nu_0) < 0,$$

and measures a contraction of the typical region depicted in Fig. 4.11; Since $E_{33} = 0$, changes in volume and area the same.

[7] In view of (2.54), the bounds $(2.44)_2$ on ν, that descend from the positivity requirement for the density of elastic energy, translate into the following equivalent bounds for ν_0:

$$-\frac{1}{2} < \nu_0 < 1.$$

Fig. 4.11 A typical area element, when polar coordinates are used

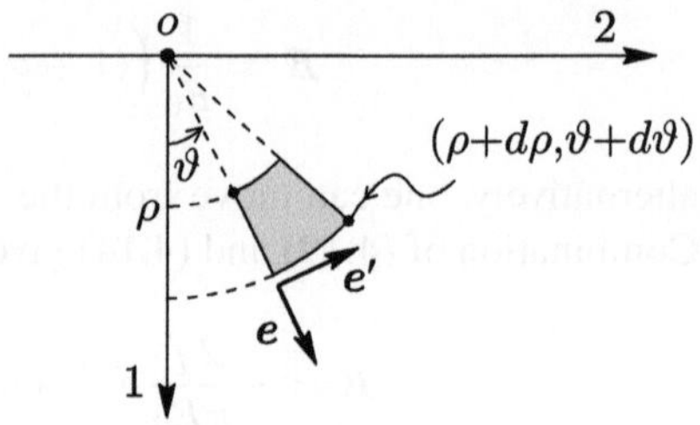

Remark 4.11 The strain field is everywhere null at the surface, that is, for $\vartheta = \pm\pi/2$. Moreover, whatever $\vartheta \in (-\pi/e, +\pi/2)$,

$$\lim_{\rho\to+\infty} \boldsymbol{E}(\rho, \vartheta) = \boldsymbol{0};$$

thus, at large distances from the origin the displacement field tends to become *rigid*.

4.5.3 The Displacement Field

The displacement field, we recall, must solve the partial differential equation

$$\nabla\boldsymbol{u} + \nabla\boldsymbol{u}^T = 2\boldsymbol{E}. \tag{4.22}$$

When general curvilinear coordinates are used,

$$\nabla\boldsymbol{u} = \boldsymbol{u},_i \otimes \boldsymbol{g}^i,$$

so that, in the present circumstances,

$$\nabla\boldsymbol{u} = \boldsymbol{u},_\rho \otimes \boldsymbol{e} + \rho^{-1}\boldsymbol{u},_\vartheta \otimes \boldsymbol{e}';$$

hence,

$$\begin{aligned} \boldsymbol{E}\boldsymbol{e}\cdot\boldsymbol{e} &= (\nabla\boldsymbol{u})\boldsymbol{e}\cdot\boldsymbol{e} = \boldsymbol{u},_\rho\cdot\boldsymbol{e} = (\boldsymbol{u}\cdot\boldsymbol{e}),_\rho\,, \\ \boldsymbol{E}\boldsymbol{e}'\cdot e' &= (\nabla\boldsymbol{u})\boldsymbol{e}'\cdot e' = \rho^{-1}\boldsymbol{u},_\vartheta\cdot e' = \rho^{-1}((\boldsymbol{u}\cdot\boldsymbol{e}'),_\vartheta + \boldsymbol{u}\cdot\boldsymbol{e}); \end{aligned} \tag{4.23}$$

in addition, we must insist that

$$\begin{aligned} 2\,\boldsymbol{E}\boldsymbol{e}\cdot\boldsymbol{e}' &= (\nabla\boldsymbol{u})\boldsymbol{e}\cdot\boldsymbol{e}' + (\nabla\boldsymbol{u})\boldsymbol{e}'\cdot\boldsymbol{e} = \boldsymbol{u},_\rho\cdot\boldsymbol{e}' + \rho^{-1}\boldsymbol{u},_\vartheta\cdot\boldsymbol{e} \\ &= (\boldsymbol{u}\cdot\boldsymbol{e}'),_\rho + \rho^{-1}\big((\boldsymbol{u}\cdot\boldsymbol{e}),_\vartheta - \boldsymbol{u}\cdot\boldsymbol{e}'\big) = 0. \end{aligned} \tag{4.24}$$

Juxtaposition of (4.20) and (4.21) with, respectively, $(4.23)_1$ and $(4.23)_2$ yields two first-order PDEs:

$$u_{\rho},_{\rho} = -\frac{2f}{\pi E_0}\rho^{-1}\cos\vartheta, \tag{4.25}$$

$$u_{\vartheta},_{\vartheta} + u_{\rho} = \nu_0\frac{2f}{\pi E_0}\cos\vartheta, \tag{4.26}$$

in the unknowns

$$u_{\rho} := \boldsymbol{u}\cdot\boldsymbol{e} = \widehat{u}_{\rho}(\rho,\vartheta), \quad u_{\vartheta} := \boldsymbol{u}\cdot\boldsymbol{e}' = \widehat{u}_{\vartheta}(\rho,\vartheta). \tag{4.27}$$

Our plan is to integrate (4.25) and (4.26), in the order, so as to find a displacement field

$$\widehat{\boldsymbol{u}}(\rho,\vartheta) = \widehat{u}_{\rho}(\rho,\vartheta)\widehat{\boldsymbol{e}}(\vartheta) + \widehat{u}_{\vartheta}(\rho,\vartheta)\widehat{\boldsymbol{e}}'(\vartheta),$$

that (i) has the necessary symmetries discussed in Sect. 4.1, that is, satisfies

$$\widehat{u}_{\rho}(\rho,\vartheta) = \widehat{u}_{\rho}(\rho,-\vartheta), \quad \widehat{u}_{\vartheta}(\rho,\vartheta) = -\widehat{u}_{\vartheta}(\rho,-\vartheta), \tag{4.28}$$

(recall relations (4.3)), and (ii) satisfies condition (4.24) as well.

Remark 4.12 The Cartesian components of the displacement field have the following expressions in terms of the physical components $(u_{\rho}, u_{\vartheta})$:

$$\begin{aligned} u_1 &= \boldsymbol{u}\cdot\boldsymbol{e}_1 = (u_{\rho}\boldsymbol{e} + u_{\vartheta}\boldsymbol{e}')\cdot\boldsymbol{e}_1 = u_{\rho}\cos\vartheta - u_{\vartheta}\sin\vartheta, \\ u_2 &= \boldsymbol{u}\cdot\boldsymbol{e}_2 = (u_{\rho}\boldsymbol{e} + u_{\vartheta}\boldsymbol{e}')\cdot\boldsymbol{e}_2 = u_{\rho}\sin\vartheta + u_{\vartheta}\cos\vartheta. \end{aligned} \tag{4.29}$$

Thus, for each ρ fixed, the first component is an even function of ϑ, and the second is odd.

Integrating Eq. (4.25) yields:

$$\widehat{u}_{\rho}(\rho,\vartheta) = -\frac{2f}{\pi E_0}\ln\rho\,\cos\vartheta + \widehat{v}(\vartheta), \tag{4.30}$$

with $\widehat{v}$ a function, to be determined later on, that must be even to obey the first of the parity conditions (4.28). With the use of this partial result, we can pass to integrating Eq. (4.26); we find:

$$\widehat{u}_{\vartheta}(\rho,\vartheta) = \frac{2f}{\pi E_0}\ln\rho\,\sin\vartheta - \widehat{V}(\vartheta) + \nu_0\frac{2f}{\pi E_0}\sin\vartheta, \tag{4.31}$$

where $\widehat{V}$ is a promitive of $\widehat{v}$, hence an odd function.

Remark 4.13 One might think that representation (4.31) for $\widehat{u}_\vartheta$ should be completed by addition of an arbitrary function of ρ. However, this would contradict condition $(4.28)_2$, according to which function $\widehat{u}_\vartheta(\rho,\cdot)$ has to be odd.

At this point, it remains for us to determine function $\widehat{v}$. We note preliminarily that, with the notations introduced in (4.27), condition (4.24) reads:

$$u_{\vartheta,\rho}+\rho^{-1}(u_{\rho,\vartheta}-u_\vartheta)=0.$$

After substitution of the expressions for u_ρ and u_ϑ given in (4.30) and (4.31), and *modulo* appropriate cancellation and rearrangement of some terms, this relation becomes:

$$\frac{2(1-\nu_0)f}{\pi E_0}\sin\vartheta+\widehat{v}'(\vartheta)+\widehat{V}(\vartheta)=0,\quad \text{with } \widehat{v} \text{ necessarily even.} \tag{4.32}$$

This simple first-order integrodifferential equation is solved by

$$\widehat{v}(\vartheta)=v_0\cos\vartheta-\frac{(1-\nu_0)f}{\pi E_0}\vartheta\sin\vartheta;$$

hence, to within an inessential constant,[8]

$$\widehat{V}(\vartheta)=v_0\sin\vartheta+\frac{(1-\nu_0)f}{\pi E_0}(\vartheta\cos\vartheta-\sin\vartheta).$$

In the light of these results, the Flamant displacement field can be given the following form:

$$\begin{aligned}\widehat{u}^F_\rho(\rho,\vartheta)&=v_0\cos\vartheta-\frac{2f}{\pi E_0}\ln\rho\cos\vartheta-\frac{(1-\nu_0)f}{\pi E_0}\vartheta\sin\vartheta,\\ \widehat{u}^F_\vartheta(\rho,\vartheta)&=-v_0\sin\vartheta+\frac{2f}{\pi E_0}\ln\rho\sin\vartheta-\frac{(1-\nu_0)f}{\pi E_0}\vartheta\cos\vartheta+\frac{(1+\nu_0)f}{\pi E_0}\sin\vartheta;\end{aligned} \tag{4.33}$$

---

[8] To find this result, (i) differentiate (4.32), and get:

$$\frac{2(1-\nu_0)f}{\pi E_0}\cos\vartheta+\widehat{v}''(\vartheta)+\widehat{v}(\vartheta)=0;$$

(ii) recall that the well-known homogeneous equation associated with this second-order ODE admits a family of even solutions:

$$\widehat{v}_h(\vartheta)=v_0\cos\vartheta;$$

(iii) confirm that function

$$\widehat{v}_p(\vartheta)=-\frac{(1-\nu_0)f}{\pi E_0}\vartheta\sin\vartheta$$

is a particular integral of the complete equation.

in Cartesian components, upon using (4.29), we have:

$$\begin{aligned}\widehat{u}_1^F(\rho,\vartheta)&=v_0-\frac{2f}{\pi E_0}\ln\rho-\frac{(1+\nu_0)f}{\pi E_0}\sin^2\vartheta,\\ \widehat{u}_2^F(\rho,\vartheta)&=-\frac{(1-\nu_0)f}{\pi E_0}\vartheta+\frac{(1+\nu_0)f}{\pi E_0}\sin\vartheta\cos\vartheta.\end{aligned}\tag{4.34}$$

Vector $\boldsymbol{v}_0=v_0\boldsymbol{e}_1$ represents an arbitrary vertical *translation* of the Flamant half-space. Such an indetermination was to be expected, in the absence of conditions bearing directly on the displacement's boundary trace. Note that the arbitrariness inherent to solving the geometric compatibility problem (4.22), which is ruled by a linear operator whose kernel is the collection of all rigid displacements, is in the present case mitigated by the imposed side requirements of symmetry: $\boldsymbol{v}_0$ is the only rigid displacement compatible with those requirements. To dispose of the residual arbitrariness by one or another choice of the constant v_0 can be regarded as a sort of completion of the boundary conditions.[9]

To have a closer look at these matters, we firstly restrict attention to the plane that limits Flamant's half-space $\mathcal{HS}^+$, where

$$\rho=|x_2|,\quad \vartheta=(\operatorname{sgn}x_2)\frac{\pi}{2};\quad u_\rho=v_0+(\operatorname{sgn}x_2)u_2,\quad u_\vartheta=-(\operatorname{sgn}x_2)u_1.$$

Then, after some manipulations of formulae (4.33) and (4.34), we find that Flamant's *surface displacement field* is:

$$\begin{aligned}\breve{u}_1(x_2)&=v_0-\frac{2f}{\pi E_0}\Big(\ln|x_2|+\frac{1}{2}(1+\nu_0)\Big),\\ \breve{u}_2(x_2)&=-\frac{(1-\nu_0)f}{2E_0}\operatorname{sgn}x_2.\end{aligned}\tag{4.35}$$

We see that, on each of the two half-planes that compose the $x_1=0$ plane, the horizontal displacement has constant value, is parallel to the x_2 axis, and is directed towards the x_3 axis: those two half-planes should then interpenetrate by a mutual sliding of $\dfrac{(1-\nu_0)f}{2E_0}$ length units.

Next, we restrict (4.33) to the x_1 axis by taking $\vartheta=0$, so that

$$\begin{aligned}\widehat{u}_1^F(\rho,0)&=v_0-\frac{2f}{\pi E_0}\ln\rho,\\ \widehat{u}_2^F(\rho,0)&\equiv 0.\end{aligned}\tag{4.36}$$

---

9 As a rule, when a boundary-value problem is formulated over an unbounded domain, the consequent lack of boundary conditions is compensated by posing on the solution a convenient set of *conditions at infinity*. This is not doable for the Flamant Problem, where the behavior at infinity of the elastic state is not tunable.

We see that the logarithmic singularity imposes that the vertical displacement be infinite both at the origin and at infinite distance from it, switching from downwards to upwards at unit distance. This last observation prompts us to choose v_0 so as to have the vertical displacement null at a point of the x_1 axis placed at a given distance ρ_0 from the origin, that is, to set

$$0 = v_0 - \frac{2f}{\pi E_0} \ln \rho_0.$$

With this measure, relations (4.33) become:

$$\begin{aligned}\widehat{u}^F_\rho(\rho, \vartheta) &= -\frac{2f}{\pi E_0} \ln \frac{\rho}{\rho_0} \cos\vartheta - \frac{(1-\nu_0)f}{\pi E_0} \vartheta \sin\vartheta, \\ \widehat{u}^F_\vartheta(\rho, \vartheta) &= -\frac{2f}{\pi E_0} \ln \frac{\rho}{\rho_0} \sin\vartheta - \frac{(1-\nu_0)f}{\pi E_0} \vartheta \cos\vartheta + \frac{(1+\nu_0)f}{\pi E_0} \sin\vartheta.\end{aligned} \tag{4.37}$$

Moreover, relations $(4.35)_1$ and $(4.36)_1$ become, respectively,

$$\breve{u}_1(x_2) = -\frac{2f}{\pi E_0}\left(\ln \frac{\rho_0}{|x_2|} + \frac{1}{2}(1+\nu_0)\right) \quad \text{and} \quad \widehat{u}^F_1(\rho, 0) = \frac{2f}{\pi E_0} \ln \frac{\rho_0}{\rho}.$$

On deciding to accept Flamant's solution only for $\rho \leq \rho_0$, we somewhat reduce the disquiet of our physical intuition, that, however, could not by any means be dissipated completely.

Remark 4.14 The incurable discrepancy between mathematical predictions and physical expectations that an analysis of the Flamant elastic state brings to light signals an intrinsic limitation of the *linear* theory of elasticity, a macroscopic theory that does not incorporate two minimal physical requirements like impenetrability and orientation preservation of matter. In *nonlinear* continuum mechanics, these two requirements are guaranteed locally by insisting that the deformation gradient $\boldsymbol{F} = \boldsymbol{I} + \nabla \boldsymbol{u}$ has a positive determinant:

$$\det \boldsymbol{F} > 0,$$

a condition that evaporates under a linearization process based on the smallness of $|\nabla \boldsymbol{u}|$ (we return on this issue in Sect. A.4 of the Appendix; see, in particular, Sect. A.4.1). This comment should not be taken as a criticism to linear elasticity, though: one should not expect any theory to have built-in applicability detectors, it is for other more encompassing theories to falsify, *à la* Popper, its predictions.

References

1. Ambrosio L, Fusco N, Pallara D (2000) Functions of bounded variation and free discontinuity problems. Oxford University Press, New York
2. Boussinesq J (1878) Équilibre d'élasticité d'un sol isotrope sans pesanteur, supportant différents poids. CR Acad Sci 86:1260–1263
3. Boussinesq J (1885) Application des Potentiels à l'Étude de l'Équilibre et du Mouvement des Solides Élastiques. Gauthiers-Villars, Paris
4. Boussinesq J (1888) Équilibre d'élasticité d'un solide sans pesanteur, homogène et isotrope, dont les parties profondes sont maintenues fixes, pendant que sa surface éprouve des pressions ou des déplacements connus, s'annullant hors d'une région restreinte où ils sont arbitraires. CR Acad Sci 106(1043–1048):1119–1123
5. Boussinesq J (1892) Des perturbations locales que produit au-dessous d'elle une forte charge, répartie uniformément le long d'une droite normale aux deux bords, à la surface supérieure d'une poutre rectangulaire et de longueur indéfinie posée de champ soit sur un sol horizontal, soit sur deux appuis transversaux équidistants de la charge. CR Acad Sci 114:1510–1516
6. Flamant A (1892) Sur la répartition des pressions dans un solide rectangulaire chargé transversalement. CR Acad Sci 114:1465–1468
7. Marzocchi A, Musesti A (2004) Balance laws and weak boundary conditions in continuum mechanics. J Elast 74(3):239–248
8. Podio-Guidugli P (2004) Examples of concentrated contact interactions in simple bodies. J Elast 75:167–186
9. Schuricht F (2007) A new mathematical foundation for contact interactions in continuum physics. Arch Rat Mech Anal 184:495–551
10. Schuricht F (2007) Interactions in continuum physics. In: Šilhavý M (ed) Mathematical modeling of bodies with complicated bulk and boundary behavior. Quaderni di Matematica 20:169–196
11. Šilhavý M (2004) Geometric integration theory and Cauchy's stress theorem. Unpublished lecture notes, March-September

Chapter 5
The Boussinesq Problem

The Boussinesq Problem (Joseph Valentin B., 1842–1929; v. [1–3]) consists in finding the elastic state in a linearly elastic isotropic half-space, subject to a concentrated load applied in a point of its boundary plane and perpendicular to it (Fig. 5.1).

Among the problems we deal with in this book, Boussinesq's is the one with the widest geotechnical applications [4, 5, 10]. As a matter of fact, a crucial feature in designing a foundation is that it transfers the service loads to the soil with a predictable and modest vertical settlement of the latter. To estimate foundation settlements by a direct use of the Boussinesq displacement solution, it would be important to assign reliable values to the elastic parameters; but, during settlement, changes are observed in the soil's mechanical response, that can compromise its bearing capability. This is why, at least for a preliminary design, the soil is generally thought of as if it were maintained in *oedometric conditions*, that is, if lateral deformations were prevented: under oedometric conditions, as we shall see, the Boussinesq vertical stress does *not* depend on the constitutive response, and can therefore be computed independently of the material response; the settlement accompanying that vertical stress is then computed by adopting more sophisticated constitutive models than isotropic linear elasticity.

It is worth noticing that a foundation is meant to spread loads, at definite variance with the concentrated force of the Boussinesq Problem. However, once the stress field induced by a concentrated load is found, the solution can be used as a Green function (recall the 1-D example given in Sect. 1.2), so as to determine the stress field induced by general load distributions; in Sect. 5.7, we exemplify how to do this.

5.1 Preliminary Symmetry Considerations

We have already observed that the 2-D versions of the Boussinesq Problem and the Flamant Problem coincide. This fact not withstanding, for the former problem dimension reduction entails a radical change of symmetries: while both the 2-D

P. Podio-Guidugli and A. Favata, *Elasticity for Geotechnicians*, Solid Mechanics and Its Applications 204, DOI: 10.1007/978-3-319-01258-2_5,

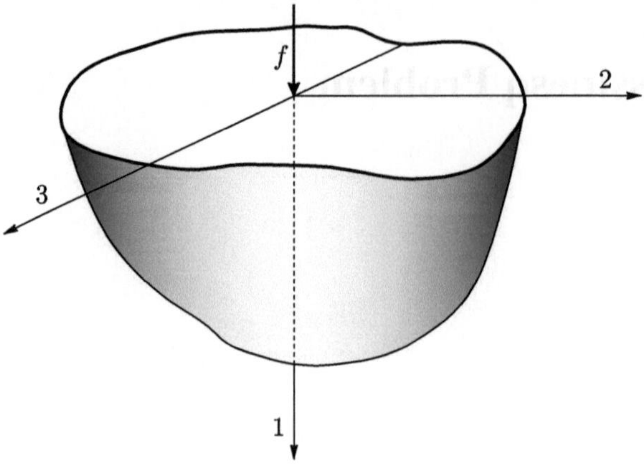

Fig. 5.1 The Boussinesq Problem

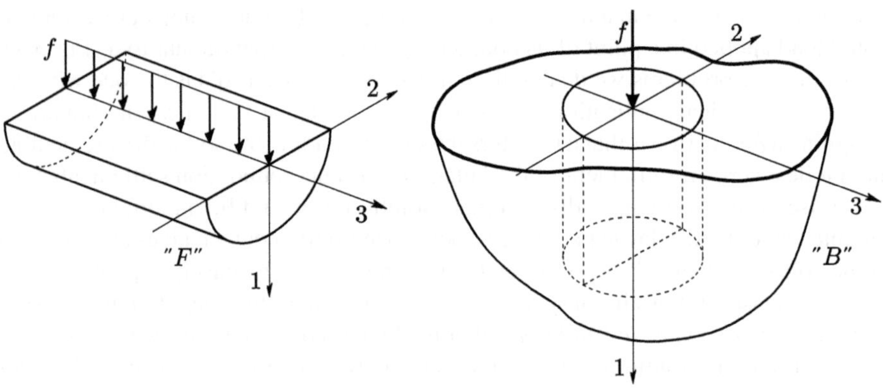

Fig. 5.2 Cylindrical symmetries in Flamant and Boussinesq Problems

Boussinesq-Flamant Problem and the Flamant Problem have cylindrical symmetry with respect to an axis parallel to $\boldsymbol{e}_3$, the Boussinesq Problem has cylindrical symmetry with respect to an axis parallel to $\boldsymbol{e}_1$ (see Fig. 5.2). Boussinesq's is a *genuinely 3-D* problem, in the sense that *it does not admits as a solution a plane displacement field*. Our attack strategy then will be different from the one used so far: firstly, we shall determine the analytic form common to all those *3-D* stress fields having the 'right' symmetries; then, we shall pick among these stress fields the only one compatible, by means of the *3-D* compatibility equation.

To recognize the right symmetries is now much less immediate than it was in the case of the Flamant Problem. In the next section, we exhibit certain stress fields that, although balanced and compliant with the boundary conditions, are not compatible. Their incompatibility descends from having symmetries that imply a counterintuitive behaviour of the accompanying deformations. The lesson to learn is that, in an elasticity problem, it is not sufficient to identify plausible *static symmetries* and reflect them into a representation of the stress fields candidate to solve the problem: only those static symmetries count that, given the body's constitution, turn out to be coherent with the *kinematic symmetries* implied by the data.

In the case of Flamant Problem, this golden rule was not evaded, even if the procedure to construct the solution left the importance of respecting the intrinsic kinematics symmetries in the background.

5.2 Inexistence of Center-Symmetric Solutions

In order to definitively convince ourselves of the importance of detecting all the relevant symmetries, we begin our study of the Boussinesq Problem by looking for a solution with plausible static symmetries, of *center-symmetric* kind. We show that there are such fields, which are balanced and satisfy both the boundary conditions and a necessary compatibility condition; and that, nevertheless, none of these fields is compatible.

5.2.1 Balanced Center-Symmetric Stress Fields

We look for divergenceless stress fields of the form:

$$\boldsymbol{S} = \sigma \boldsymbol{r} \otimes \boldsymbol{r}, \quad \sigma = \widehat{\sigma}(\rho, \vartheta), \quad \text{with } \widehat{\sigma}(\rho, \vartheta) = \widehat{\sigma}(\rho, -\vartheta). \tag{5.1}$$

For such stress fields, the stress vector is radial at any point of any hemisphere $\frac{1}{2}\mathcal{S}_\rho$ (Fig. 5.3), where the outward normal $\boldsymbol{n} = \boldsymbol{r}$ and hence

$$\boldsymbol{S}\boldsymbol{n} = (\sigma \boldsymbol{r} \otimes \boldsymbol{r})\boldsymbol{r} = \sigma \boldsymbol{r};$$

moreover, the stress vector is null at all points of the boundary plane (where $\boldsymbol{r} = \boldsymbol{h}$ and $\boldsymbol{n} = -\boldsymbol{e}_1$, and hence $\boldsymbol{S}\boldsymbol{n} = \boldsymbol{0}$), except at the origin, where, however, it is not susceptible of a classical definition, due to the unavoidable singularity of the stress field.

That the stress fields (5.1) should be singular at the origin is indicated by laying down the requirement that the integral over $\frac{1}{2}\mathcal{S}_\rho$ of the stress vector balances the applied load $\boldsymbol{f} = f\boldsymbol{e}_1$:

$$\boldsymbol{0} = f\boldsymbol{e}_1 + \int_{\frac{1}{2}\mathcal{S}_\rho} \widehat{\sigma}(\rho, \vartheta)\widehat{\boldsymbol{r}}(\vartheta, \varphi)\rho^2 |\sin\vartheta|\, d\vartheta d\varphi. \tag{5.2}$$

On recalling (3.13), we see that this vectorial equation is in fact equivalent to the scalar equation

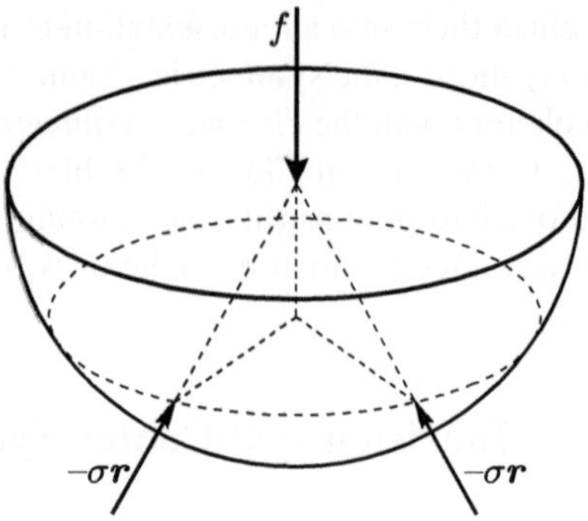

Fig. 5.3 Equilibrium of a hemispherical part

$$f+\pi\rho^2\Big(\int_{-\pi/2}^{+\pi/2}\widehat{\sigma}(\rho,\vartheta)\cos\vartheta|\sin\vartheta|\,d\vartheta\Big)=0,\quad \text{for all } \rho>0. \tag{5.3}$$

Now, for

$$\lim_{\rho\to+\infty}\rho^2\Big(\int_{-\pi/2}^{+\pi/2}\widehat{\sigma}(\rho,\vartheta)\cos\vartheta|\sin\vartheta|\,d\vartheta\Big)$$

to be finite, as required by (5.2), it is necessary that

$$\widehat{\sigma}(\rho,\vartheta)=\rho^{-2}\widehat{\tau}(\vartheta)+o(\rho^{-2}).$$

Note that the singularity in the stress field must have 'strength' 2, as a consequence of the genuine three-dimensionality of the problem at hand (the strength is 1 in Flamant's case, as is in the case of the Cerruti Problem to be studied in Chap. 8, because those two problems are not genuinely three-dimensional, in that they admit a plane displacement solution).

To validate and sharpen this prediction, recall that the divergence operator in curvilinear coordinates has the general expression (3.17). Combining Eqs. (3.12) and (3.17) with Eq. (5.1), we find that[1]

$$\operatorname{div}\boldsymbol{S}=(\sigma,_\rho+2\rho^{-1}\sigma)\boldsymbol{r} \tag{5.4}$$

---

[1] The information items needed for this calculation are:

$$\boldsymbol{S}\boldsymbol{g}^1=\sigma\boldsymbol{r},\quad \boldsymbol{S}\boldsymbol{g}^2=\boldsymbol{S}\boldsymbol{g}^3=\boldsymbol{0},\quad \boldsymbol{S}\boldsymbol{h}=(\sin\vartheta)\sigma\boldsymbol{r};\quad \boldsymbol{r},_{\vartheta\vartheta}=-\boldsymbol{r},\quad \boldsymbol{r},_{\varphi\varphi}=-(\sin\vartheta)\boldsymbol{h}.$$

With this, one finds:

$$\begin{aligned}\boldsymbol{S},_\rho\boldsymbol{g}^1+\boldsymbol{S},_\vartheta\boldsymbol{g}^2+\boldsymbol{S},_\varphi\boldsymbol{g}^3&=(\boldsymbol{S}\boldsymbol{g}^1),_\rho+(\boldsymbol{S}\boldsymbol{g}^2),_\vartheta-\rho^{-1}\boldsymbol{S}\boldsymbol{r},_{\vartheta\vartheta}+(\boldsymbol{S}\boldsymbol{g}^3),_\varphi-(\rho\sin^2\vartheta)^{-1}\boldsymbol{S}\boldsymbol{r},_{\varphi\varphi}\\&=(\sigma\boldsymbol{r}),_\rho+\rho^{-1}\sigma\boldsymbol{r}+(\rho\sin^2\vartheta)^{-1}(\sin\vartheta)\boldsymbol{S}\boldsymbol{h},\end{aligned}$$

whence (5.4) easily follows.

Hence, for a stress field of the form (5.1) to be divergenceless, the scalar field σ must satisfy

$$\sigma,_{\rho} + 2\rho^{-1}\sigma = 0, \tag{5.5}$$

and hence have the form[2]

$$\widehat{\sigma} = \rho^{-2}\widehat{\tau}(\vartheta), \quad \text{with} \quad \widehat{\tau}(\vartheta) = \widehat{\tau}(-\vartheta).$$

In conclusion, we have determined a whole family of center-symmetric stress fields balancing the given boundary load, of the form:

$$\widehat{\boldsymbol{S}}(\rho, \vartheta, \varphi) = \rho^{-2}\widehat{\tau}(\vartheta)\widehat{\boldsymbol{r}}(\vartheta, \varphi) \otimes \widehat{\boldsymbol{r}}(\vartheta, \varphi); \tag{5.6}$$

such family is parameterized by the even function $\widehat{\tau}$.

5.2.2 Incompatibility of Center-Symmetric Stress Fields

To solve Boussinesq's problem, the function $\widehat{\tau}$ must be chosen so as to satisfy the *compatibility condition in stress terms*. For the homogeneous and isotropic linearly elastic materials contemplated by classical elasticity, when the distance loads are null, this condition has the form (2.68), reproduced here for convenience:

$$\Delta \boldsymbol{S} + \frac{1}{1+\nu}\nabla\nabla(\operatorname{tr}\boldsymbol{S}) = \boldsymbol{0};$$

a general consequence of (2.68) is relation (2.67):

$$\Delta(\operatorname{tr}\boldsymbol{S}) = 0.$$

For stress fields of the form (5.6), (2.67) reads:

$$\Delta(\rho^{-2}\widehat{\tau}(\vartheta)) = 0.$$

It is not difficult to give this equation the form:

$$\sin\vartheta\,\tau'' + \cos\vartheta\,\tau' + 2\sin\vartheta\,\tau = 0, \tag{5.7}$$

[2] On differentiating (5.3) with respect to ρ, we quickly find that

$$\int_{-\pi/2}^{+\pi/2} \big(2\widehat{\sigma}(\rho, \vartheta) + \rho\widehat{\sigma},_{\rho}(\rho, \vartheta)\big)\cos\vartheta\, d\vartheta = 0.$$

We are then driven to choose a mapping $\widehat{\sigma}$ that satisfies the partial differential equation (5.5).

nor it is difficult to show that the only *regular* fundamental solution of this homogeneous second-order ODE is[3]:

$$\widehat{\tau}(\vartheta) = \tau_0 \cos\vartheta. \tag{5.8}$$

Unfortunately, the only stress field of the form

$$\mathbf{S} = \tau_0 \widetilde{\mathbf{S}}, \quad \widetilde{\mathbf{S}} := \rho^{-2}\cos\vartheta\, \boldsymbol{r}\otimes\boldsymbol{r} \tag{5.9}$$

which solves (2.68) is the uninteresting everywhere null field.[4] In the next section, we turn our attention to stress fields with cylindrical symmetry about an axis parallel to $\boldsymbol{e}_1$.

---

[3] Point $\vartheta = 0$ is the only one in the interval $(-\pi/2, +\pi/2)$ where Eq. (5.7) is singular. The other fundamental solution of this equation being singular at that point is:

$$\check{\tau}(\vartheta) = 1 + \cos\vartheta \log\tan\frac{\vartheta}{2}.$$

Here is a method to construct this solution. It is not difficult to show that (5.7) is equivalent to

$$(\sin\vartheta\, W(\vartheta))' = 0,$$

where

$$W(\vartheta) = \cos\vartheta\, \check{\tau}'(\vartheta) + \sin\vartheta\, \check{\tau}(\vartheta)$$

is the wronskian of $\widehat{\tau}$ and $\check{\tau}$. Hence, *modulo* a constant,

$$W(\vartheta) = \frac{1}{\sin\vartheta},$$

and the combination of the last two relations yields the following first order ODE for $\check{\tau}$:

$$\check{\tau}' + \frac{\sin\vartheta}{\cos\vartheta}\check{\tau} = \frac{1}{\sin\vartheta\cos\vartheta},$$

which can be re-written in the form

$$\left(\frac{1}{\cos\vartheta}\check{\tau}\right)' = \frac{1}{\sin\vartheta\cos^2\vartheta}.$$

The last bit of information needed is:

$$\int \frac{1}{\sin\vartheta\cos^2\vartheta} = \frac{1}{\cos\vartheta} + \log\tan\frac{\vartheta}{2}.$$

[4] In an attempt to satisfy (2.68) with the field (5.9), it is found that

$$\Delta\widetilde{\mathbf{S}} + \frac{1}{1+\nu}\nabla\nabla(\operatorname{tr}\widetilde{\mathbf{S}}) \neq \boldsymbol{0}.$$

5.3 Balanced Stress Fields with Cylindrical Symmetry

By a *stress field with* $\boldsymbol{e}_1$*-cylindrical symmetry* we mean a stress field whose physical components are all independent of the angle φ.

Consider a cylindrical part of the Boussinesq half-space, of arbitrary height z and radius r (Fig. 5.4). The rotational balance of such a part about the vertical axis involves the stress components $S_{\varphi r} := (\boldsymbol{S}\boldsymbol{h}) \cdot \boldsymbol{h}'$ and $S_{\varphi z} := (\boldsymbol{S}\boldsymbol{e}_1) \cdot \boldsymbol{h}'$ only; since the latter are both assumed to be independent of φ, that balance reads as follows:

$$\int_0^r t^2 S_{\varphi z}(z, t)\, dt + r^2 \int_0^z S_{\varphi r}(s, r)\, ds = 0, \quad \text{for all positive } r, z. \tag{5.10}$$

We shall now show that this balance condition is satisfied, because both $S_{\varphi r}$ and $S_{\varphi z}$ are null, due to the symmetries that a displacement field solving the Boussinesq problem must have.

Our physical intuition tells us that a *displacement field with cylindrical symmetry* is such that

$$u_\varphi \equiv 0 \quad \text{and, in addition, both } u_z \text{ and } u_r \text{ are independent of } \varphi; \tag{5.11}$$

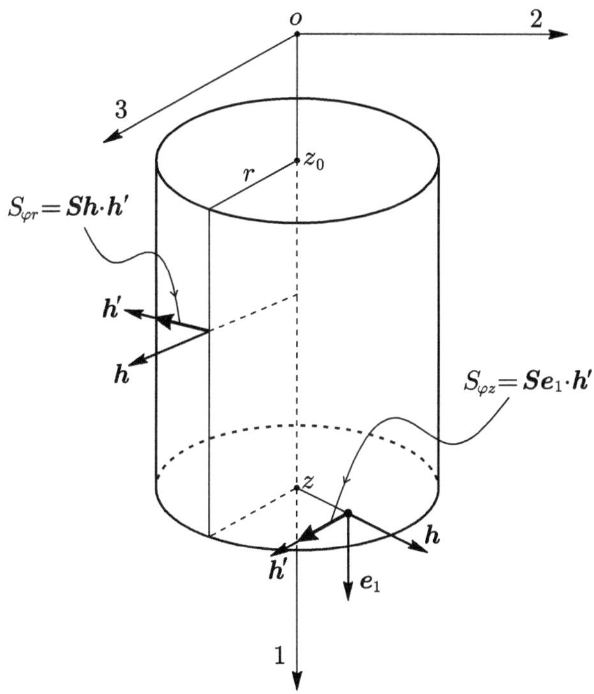

Fig. 5.4 Rotational equilibrium about the vertical axis of a cylindrical part

such a priori restrictions on the representation of $\boldsymbol{u}$ are precisely those accepted when a solution to the Boussinesq Problem is sought. On the other hand, in view of the definition of the strain measure $\boldsymbol{E}$, we have that

$$\begin{aligned} 2\,E_{\varphi r} &= ((\nabla \boldsymbol{u})\boldsymbol{h}) \cdot \boldsymbol{h}' + ((\nabla \boldsymbol{u})\boldsymbol{h}') \cdot \boldsymbol{h} = \boldsymbol{u},_r \cdot \boldsymbol{h}' + r^{-1}\boldsymbol{u},_\varphi \cdot \boldsymbol{h} \\ &= u_\varphi,_r + r^{-1}(u_r,_\varphi - u_\varphi) \end{aligned}$$

and

$$\begin{aligned} 2\,E_{\varphi z} &= ((\nabla \boldsymbol{u})\boldsymbol{e}_1) \cdot \boldsymbol{h}' + ((\nabla \boldsymbol{u})\boldsymbol{h}') \cdot \boldsymbol{e}_1 = \boldsymbol{u},_z \cdot \boldsymbol{h}' + r^{-1}\boldsymbol{u},_\varphi \cdot \boldsymbol{e}_1 \\ &= u_\varphi,_z + r^{-1}u_z,_\varphi, \end{aligned}$$

whatever the symmetries of the field $\boldsymbol{u}$.[5] Now, it is not difficult to see that both strain components $E_{\varphi r}$ and $E_{\varphi z}$ are null if (5.11) prevail. But then, given that the Boussinesq half-space is comprised of an isotropic, linearly elastic material, both stress components $S_{\varphi r}$ and $S_{\varphi z}$ must also be null; hence, in particular, the balance condition (5.10) is satisfied.

The two conditions:

$$S_{\varphi r} \equiv 0 \quad \text{and} \quad S_{\varphi z} \equiv 0 \quad \text{in } \mathcal{HS}^+$$

add to the purely static condition that a stress field solving Boussinesq Problem should have $\boldsymbol{e}_1$-cylindrical symmetry. Our first conclusion is that, with the use of the physical basis (3.8), the class of stress fields of interest has the representation:

$$\boldsymbol{S} = \sigma_1 \boldsymbol{e}_1 \otimes \boldsymbol{e}_1 + \sigma_2 \boldsymbol{h} \otimes \boldsymbol{h} + \sigma_3 \boldsymbol{h}' \otimes \boldsymbol{h}' + \sigma_4(\boldsymbol{e}_1 \otimes \boldsymbol{h} + \boldsymbol{h} \otimes \boldsymbol{e}_1), \tag{5.12}$$

parameterized by the four scalar-valued mappings $\widehat{\sigma}_i$, with

$$\begin{aligned} \widehat{\sigma}_1(z,r) &= S_{zz} = \boldsymbol{S} \cdot \boldsymbol{e}_1 \otimes \boldsymbol{e}_1, \\ \widehat{\sigma}_2(z,r) &= S_{rr} = \boldsymbol{S} \cdot \boldsymbol{h} \otimes \boldsymbol{h}, \\ \widehat{\sigma}_3(z,r) &= S_{\varphi\varphi} = \boldsymbol{S} \cdot \boldsymbol{h}' \otimes \boldsymbol{h}', \\ \widehat{\sigma}_4(z,r) &= S_{zr} = \boldsymbol{S} \cdot \boldsymbol{e}_1 \otimes \boldsymbol{h}. \end{aligned}$$

Given that, in cylindrical coordinates,

---

[5] To obtain the last two relations, it is useful to recall that

$$\nabla \boldsymbol{u} = \boldsymbol{u},_z \otimes \boldsymbol{e}_1 + \boldsymbol{u},_r \otimes \boldsymbol{h} + r^{-1}\boldsymbol{u},_\varphi \otimes \boldsymbol{h}',$$

and that the physical components of $\boldsymbol{u}$ are:

$$u_z := \boldsymbol{u} \cdot \boldsymbol{e}_1, \quad u_r := \boldsymbol{u} \cdot \boldsymbol{h}, \quad u_\varphi := \boldsymbol{u} \cdot \boldsymbol{h}'.$$

$$\operatorname{div} \boldsymbol{S} = (\boldsymbol{S}\boldsymbol{e}_1)_{,z} + (\boldsymbol{S}\boldsymbol{h})_{,r} + r^{-1}(\boldsymbol{S}\boldsymbol{h}')_{,\varphi} + r^{-1}\boldsymbol{S}\boldsymbol{h},$$

a stress field of type (5.12) is balanced for null distance forces if

$$\begin{aligned}\boldsymbol{0} &= r(\sigma_1\boldsymbol{e}_1 + \sigma_4\boldsymbol{h})_{,z} + \big(r(\sigma_2\boldsymbol{h} + \sigma_4\boldsymbol{e}_1)\big)_{,r} + (\sigma_3\boldsymbol{h}')_{,\varphi}\\ &= \big(r\sigma_{1,z} + (r\sigma_4)_{,r}\big)\boldsymbol{e}_1 + \big(r\sigma_{4,z} + (r\sigma_2)_{,r} - \sigma_3\big)\boldsymbol{h}.\end{aligned}$$

The choice of parameter mappings is therefore restricted to those obeying the following partial differential equations:

$$r\sigma_{1,z} + (r\sigma_4)_{,r} = 0, \tag{5.13}$$

$$r\sigma_{4,z} + (r\sigma_2)_{,r} - \sigma_3 = 0. \tag{5.14}$$

5.3.1 Boundary Conditions

A further restriction on the choice of parameter mappings comes from the need to satisfy the boundary condition of null traction at all points of the plane $z = 0$, *origin excluded.* This condition is:

$$\widehat{\boldsymbol{S}}(0, r, \varphi)\boldsymbol{e}_1 = \boldsymbol{0}, \quad r > 0, \tag{5.15}$$

or rather, in terms of the representation (5.12),

$$\widehat{\sigma}_1(0, r) = \widehat{\sigma}_4(0, r) = 0, \quad r > 0. \tag{5.16}$$

The boundary condition prevailing on the *whole* plane $z = 0$ may be written in the weak form

$$\widetilde{\boldsymbol{S}}(x)\boldsymbol{e}_1 = \delta(x - o) f \boldsymbol{e}_1 \quad \text{for all } x \text{ such that } (x - o)\cdot\boldsymbol{e}_1 = 0. \tag{5.17}$$

In addition to (5.11), another symmetry condition prevails, namely,

$$\lim_{r\to 0^+} u_r(z, r) = 0;$$

as we shall see later on, this last condition works in all respects as a boundary conditions.

5.3.2 A Kinematic Condition

A consequence of the symmetry condition (5.11) we have not yet discussed is that

$$\nabla\boldsymbol{u} = u_{z,z}\,\boldsymbol{e}_1\otimes\boldsymbol{e}_1 + u_{r,z}\,\boldsymbol{h}\otimes\boldsymbol{e}_1 + r^{-1}u_r\boldsymbol{h}'\otimes\boldsymbol{h}' + u_{z,r}\,\boldsymbol{e}_1\otimes\boldsymbol{h} + u_{r,r}\,\boldsymbol{h}\otimes\boldsymbol{h};$$

it follows that

$$E_{zz} = u_{z},_{z}, E_{rr} = u_{r},_{r}, E_{\varphi\varphi} = r^{-1}u_r, \quad E_{zr} = \frac{1}{2}\Big(u_{z},_{r} + u_{r},_{z}\Big),$$

whence, in particular, that

$$E_{rr} = (rE_{\varphi\varphi}),_{r}. \tag{5.18}$$

In view of the inverse constitutive equation (2.45), this consistency condition of representation (5.11) can be rewritten as:

$$\sigma_2 - (r\sigma_3),_{r} + \nu r\alpha,_{r} = 0. \tag{5.19}$$

Condition (5.19) is related to the compatibility equation

$$\operatorname{curl}\operatorname{curl}\boldsymbol{E} = \boldsymbol{0}.$$

In fact, given the noted symmetries in the displacement and deformation fields, it is not difficult to see that the $(\boldsymbol{e}_1 \otimes \boldsymbol{e}_1)$-component of $\operatorname{curl}\operatorname{curl}\boldsymbol{E}$ is given by:

$$\operatorname{curl}\operatorname{curl}\boldsymbol{E}\cdot\boldsymbol{e}_1 \otimes \boldsymbol{e}_1 =: (\operatorname{curl}\operatorname{curl}\boldsymbol{E})_{zz} = 0 = r^{-1}E_{rr,r} - E_{\varphi\varphi,rr} - 2r^{-1}E_{\varphi\varphi,r},$$

or rather, equivalently,

$$E_{rr,r} = (rE_{\varphi\varphi}),_{rr},$$

a direct consequence of (5.18).

Remark 5.1 It is not difficult to obtain the following balance of the contact actions on a cylindrical part, whose axis is the coordinate axis $x_2 = x_3 = 0$, of radius r and height $(z - z_0)$:

$$\int_0^r t\big(\widehat{\sigma}_1(z,t) - \widehat{\sigma}_1(z_0,t)\big)dt + r\int_{z_0}^{z} \widehat{\sigma}_4(s,r)ds = 0, \quad \text{per ogni } z > z_0 > 0, r > 0. \tag{5.20}$$

On differentiating this relation with respect to r, we obtain

$$r\big(\widehat{\sigma}_1(z,r) - \widehat{\sigma}_1(0,r)\big) + \int_{z_0}^{z} \widehat{\sigma}_4(s,r)ds + r\int_{z_0}^{z} \widehat{\sigma}_{4},_{r}(s,r)ds = 0;$$

a further differentiation, this time with respect to z, yields:

$$r\sigma_{1},_{z} + \sigma_4 + r\sigma_{4},_{r} = 0,$$

that is the differential equilibrium condition (5.13). On the other hand, it is obvious that it is possible to obtain (5.20), a *part-wise* condition, by integrating (5.13), a *local* condition valid in each point of $\mathcal{HS}^+$. Not surprisingly, we conclude that both

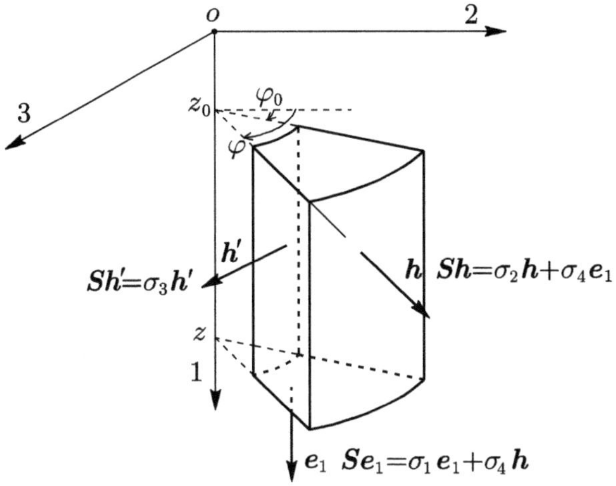

Fig. 5.5 Radial equilibrium of a portion of cylindrical sector

relations have coherent mechanical interpretation, but different cogency. As a matter of fact, at equilibrium it has to be null the vertical component of the resultant of the actions exerted on: an arbitrary part, in the case of (5.13); a cylindrical part of the indicated kind, in the case of (5.20).

Remark 5.2 To find the mechanical interpretation of the equilibrium relation (5.14) is also not too difficult a task: one differentiates the condition expressing the balance of contact actions on the part of cylindrical sector shown in Fig. 5.5: and observes that the outcome prescribes that the radial actions exerted on the surface of that sector sum up to null at equilibrium. Indeed, the resultants of those contact actions are:

(i) on faces parallel to the coordinate plane $x_1 = 0$,

$$\int_{r_0}^{r}\int_{\varphi_0}^{\varphi}\Big(\big(\widehat{\sigma}_1(z,s)-\widehat{\sigma}_1(z_0,s)\big)\boldsymbol{e}_1+\big(\widehat{\sigma}_4(z,s)+\widehat{\sigma}_4(z_0,s)\big)\boldsymbol{h}(\psi)\Big)s\,ds\,d\psi;$$

(ii) on vertical faces,

$$\int_{z_0}^{z}\int_{r_0}^{r}\big(\widehat{\sigma}_3(t,s)\boldsymbol{h}'(\varphi)-\widehat{\sigma}_3(t,s)\boldsymbol{h}'(\varphi_0)\big)dt\,ds;$$

(iii) on the two remaining faces,

$$\int_{z_0}^{z}\int_{\varphi_0}^{\varphi}\Big(r\big(\widehat{\sigma}_2(t,r)\boldsymbol{h}(\psi)+\widehat{\sigma}_4(t,r)\boldsymbol{e}_1\big)+r_0\big(\widehat{\sigma}_2(t,r_0)\boldsymbol{h}(\psi)+\widehat{\sigma}_4(t,r_0)\boldsymbol{e}_1\big)\Big)\,dt\,d\psi.$$

Summing up and differentiating with respect to φ, we obtain the vector equation:

$$\Big(\int_{r_0}^{r}(\widehat{\sigma}_4(z,s)+\widehat{\sigma}_4(z_0,s))s\,ds\Big)\boldsymbol{h}(\varphi)-\Big(\int_{z_0}^{z}\int_{r_0}^{r}\widehat{\sigma}_3(t,s)\,dt\,ds\Big)\boldsymbol{h}(\varphi)$$
$$+\Big(\int_{z_0}^{z}(r\widehat{\sigma}_2(t,r)+r_0\widehat{\sigma}_2(t,r_0))dt\Big)\boldsymbol{h}(\varphi)=\boldsymbol{0},$$

from which, this time differentiating with respect to r, it follows that:

$$r\widehat{\sigma}_4(z,r)-\int_{z_0}^{z}\widehat{\sigma}_3(t,r)\,dt+\int_{z_0}^{z}(r\widehat{\sigma}_2(t,r))_{,r}\,dt=0;$$

finally, on differentiating with respect to z, we reproduce (5.14).

Remark 5.3 We collect here some formulas of repeated use in what follows.
Let $\alpha=\widehat{\alpha}(z,r)$ be a scalar field of class C^2. Then,

$$\nabla\alpha=\alpha_{,z}\,\boldsymbol{e}_1+\alpha_{,r}\,\boldsymbol{h}; \tag{5.21}$$
$$\nabla^2\alpha=\alpha_{,zz}\,\boldsymbol{e}_1\otimes\boldsymbol{e}_1+\alpha_{,rr}\,\boldsymbol{h}\otimes\boldsymbol{h}+r^{-1}\alpha_{,r}\,\boldsymbol{h}'\otimes\boldsymbol{h}'$$
$$+\,\alpha_{,zr}\,(\boldsymbol{e}_1\otimes\boldsymbol{h}+\boldsymbol{h}\otimes\boldsymbol{e}_1); \tag{5.22}$$
$$\Delta\alpha=\alpha_{,zz}+\alpha_{,rr}+r^{-1}\alpha_{,r}\,. \tag{5.23}$$

We also record, again in view of a later use, a consequence of definition (5.23):

$$\Delta(r\alpha)=r\Delta\alpha+2\alpha_{,r}+r^{-1}\alpha. \tag{5.24}$$

Remark 5.4 From relations (5.13) and (5.14) a differential condition follows, involving functions $\widehat{\sigma}_1,\widehat{\sigma}_2$ e $\widehat{\sigma}_3$ and to be useful in the following. Here it is.
Given that an application of (5.21) yields:

$$\nabla(r\sigma_4)=-((r\sigma_2)_{,r}-\sigma_3)\boldsymbol{e}_1-r\sigma_{1,z}\,\boldsymbol{h}, \tag{5.25}$$

we have that

$$-\nabla^2(r\sigma_4)=((r\sigma_2)_{,r}-\sigma_3)_{,z}\,\boldsymbol{e}_1\otimes\boldsymbol{e}_1+r\sigma_{1,zz}\,\boldsymbol{h}\otimes\boldsymbol{e}_1$$
$$+((r\sigma_2)_{,r}-\sigma_3)_{,r}\,\boldsymbol{e}_1\otimes\boldsymbol{h}+(r\sigma_{1,z})_{,r}\,\boldsymbol{h}\otimes\boldsymbol{h}+\sigma_{1,z}\,\boldsymbol{h}'\otimes\boldsymbol{h}'. \tag{5.26}$$

To guarantee the indispensable symmetry of this second gradient (or, if one so prefers, to guarantee that curl $\nabla(r\sigma_4)=\boldsymbol{0}$), functions $\widehat{\sigma}_i,\ i=1,2,3$ must fulfill the following condition[6]:

---

[6] For an alternative way to deduce this condition, one writes (5.13) and (5.14) in the form:

$$(r\sigma_4)_{,r}=-r\sigma_1,z,\quad (r\sigma_4),z=\sigma_3-(r\sigma_2)_{,r}\,;$$

$$r\sigma_1{,}_{zz} - (r\sigma_2){,}_{rr} + \sigma_3{,}_r = 0.$$

Remark 5.5 A further consequence of (5.26) is that

$$\nabla^2(r\sigma_4)\cdot \boldsymbol{I} = \Delta(r\sigma_4) = -r(\sigma_1+\sigma_2){,}_{zr} - (2\sigma_1+\sigma_2-\sigma_3){,}_z\,.$$

On the other hand, if (5.24) is written for $\alpha = \sigma_4$, it is found that:

$$\Delta(r\sigma_4) = r\Delta\sigma_4 + 2\sigma_4{,}_r + r^{-1}\sigma_4.$$

A combination of the last two relations gives[7]:

$$\Delta\sigma_4 + 2r^{-1}\sigma_4{,}_r + r^{-2}\sigma_4 = -(\sigma_1+\sigma_2){,}_{zr} - r^{-1}(2\sigma_1+\sigma_2-\sigma_3){,}_z\,,$$

$$\Delta\sigma_4 - r^{-2}\sigma_4 = -(\sigma_1+\ \sigma_2+\sigma_3){,}_{zr} - r^{-1}(\sigma_2-(r\sigma_3){,}_r\,){,}_z\,. \tag{5.27}$$

5.3.3 Balanced and Compatible Stress Fields

First of all, we record two useful differential identities, holding for sufficiently smooth but otherwise arbitrary scalar and vector fields. The first identity is:

$$\begin{aligned}\Delta(\alpha\boldsymbol{a}\otimes\boldsymbol{a}) &= (\Delta\alpha)\boldsymbol{a}\otimes\boldsymbol{a} + 2(\boldsymbol{a}\otimes(\nabla\boldsymbol{a}(\nabla\alpha)) + (\nabla\boldsymbol{a}(\nabla\alpha))\otimes\boldsymbol{a})\\ &\quad+\alpha(\boldsymbol{a}\otimes\Delta\boldsymbol{a} + \Delta\boldsymbol{a}\otimes\boldsymbol{a} + 2\nabla\boldsymbol{a}\nabla\boldsymbol{a}^T);\end{aligned} \tag{5.28}$$

the second is:

$$\begin{aligned}\Delta(\alpha(\boldsymbol{a}\otimes\boldsymbol{b}+\boldsymbol{b}\otimes\boldsymbol{a})) &= (\Delta\alpha)(\boldsymbol{a}\otimes\boldsymbol{b}+\boldsymbol{b}\otimes\boldsymbol{a}) + 2(\boldsymbol{b}\otimes(\nabla\boldsymbol{a}(\nabla\alpha))\\ &\quad+(\nabla\boldsymbol{a}(\nabla\alpha))\otimes\boldsymbol{b}) + \alpha(\boldsymbol{b}\otimes\Delta\boldsymbol{a} + \Delta\boldsymbol{a}\otimes\boldsymbol{b}),\end{aligned} \tag{5.29}$$

where $\boldsymbol{b}$ is a constant-valued vector field. It is not difficult to see that these two identities follow from the more general identity:

$$\begin{aligned}\Delta(\alpha\boldsymbol{a}\otimes\boldsymbol{b}) &= (\Delta\alpha)\boldsymbol{a}\otimes\boldsymbol{b} + 2(\boldsymbol{a}\otimes(\nabla\boldsymbol{b}(\nabla\alpha)) + (\nabla\boldsymbol{a}(\nabla\alpha))\otimes\boldsymbol{b})\\ &\quad+\alpha(\Delta\boldsymbol{a}\otimes\boldsymbol{b} + 2\nabla\boldsymbol{a}\nabla\boldsymbol{b}^T + \boldsymbol{a}\otimes\Delta\boldsymbol{b}).\end{aligned}$$

(Footnote 6 continued)
differentiates the first equation with respect to z, the second with respect to r: and finishes by eliminating $(r\sigma_4){,}_{zr}$.

[7] To take the last step in the calculation, use has been made of the following alternative version of (5.13):

$$r^{-1}\sigma_4{,}_r + r^{-2}\sigma_4 = -r^{-1}\sigma_1,z.$$

With a view toward determining the implications of the tensorial compatibility condition (2.68) on the choice of the functions $\widehat{\sigma}_i$ in a representation of type (5.12) for the stress field, we observe that

$$\Delta \boldsymbol{S} = \Delta(\sigma_1 \boldsymbol{e}_1 \otimes \boldsymbol{e}_1) + \Delta(\sigma_2 \boldsymbol{h} \otimes \boldsymbol{h}) + \Delta(\sigma_3 \boldsymbol{h}' \otimes \boldsymbol{h}') + \Delta(\sigma_4(\boldsymbol{e}_1 \otimes \boldsymbol{h} + \boldsymbol{h} \otimes \boldsymbol{e}_1)).$$

On applying repeatedly first (5.28) and then (5.29), and on taking into account the following properties of the fields $\boldsymbol{h}$ and $\boldsymbol{h}'$:

$$\begin{aligned} \nabla \boldsymbol{h} = r^{-1}\boldsymbol{h}' \otimes \boldsymbol{h}', \quad & \Delta \boldsymbol{h} = -r^{-2}\boldsymbol{h}; \\ \nabla \boldsymbol{h}' = r^{-1}\boldsymbol{h} \otimes \boldsymbol{h}', \quad & \Delta \boldsymbol{h}' = -r^{-2}\boldsymbol{h}'. \end{aligned}$$

we find:

$$\begin{aligned} \Delta(\sigma_1 \boldsymbol{e}_1 \otimes \boldsymbol{e}_1) &= (\Delta\sigma_1)\boldsymbol{e}_1 \otimes \boldsymbol{e}_1; \\ \Delta(\sigma_2 \boldsymbol{h} \otimes \boldsymbol{h}) &= (\Delta\sigma_2)\boldsymbol{h} \otimes \boldsymbol{h} - 2r^{-2}\sigma_2(\boldsymbol{h} \otimes \boldsymbol{h} - \boldsymbol{h}' \otimes \boldsymbol{h}'); \\ \Delta(\sigma_3 \boldsymbol{h}' \otimes \boldsymbol{h}') &= (\Delta\sigma_3)\boldsymbol{h}' \otimes \boldsymbol{h}' + 2r^{-2}\sigma_3(\boldsymbol{h} \otimes \boldsymbol{h} - \boldsymbol{h}' \otimes \boldsymbol{h}'); \\ \Delta(\sigma_4(\boldsymbol{e}_1 \otimes \boldsymbol{h} + \boldsymbol{h} \otimes \boldsymbol{e}_1) &= (\Delta\sigma_4 - r^{-2}\sigma_4)(\boldsymbol{e}_1 \otimes \boldsymbol{h} + \boldsymbol{h} \otimes \boldsymbol{e}_1). \end{aligned}$$

Then,

$$\begin{aligned} \Delta \boldsymbol{S} = {} & (\Delta\sigma_1)\boldsymbol{e}_1 \otimes \boldsymbol{e}_1 + \big(\Delta\sigma_2 - 2r^{-2}(\sigma_2 - \sigma_3)\big)\boldsymbol{h} \otimes \boldsymbol{h} \\ & + \big(\Delta\sigma_3 + 2r^{-2}(\sigma_2 - \sigma_3)\big)\boldsymbol{h}' \otimes \boldsymbol{h}' + (\Delta\sigma_4 - r^{-2}\sigma_4)(\boldsymbol{e}_1 \otimes \boldsymbol{h} + \boldsymbol{h} \otimes \boldsymbol{e}_1). \end{aligned}$$

Thus, on applying (5.22) for

$$\alpha := (1+\nu)^{-1}\mathrm{tr}\, S = (1+\nu)^{-1}(\sigma_1 + \sigma_2 + \sigma_3), \tag{5.30}$$

we have that

$$\begin{aligned} \Delta \boldsymbol{S} + \frac{1}{1+\nu}\nabla^2(\mathrm{tr}\,\boldsymbol{S}) = {} & (\Delta\sigma_1)\boldsymbol{e}_1 \otimes \boldsymbol{e}_1 + (\Delta\sigma_2 - 2r^{-2}(\sigma_2 - \sigma_3)\boldsymbol{h} \otimes \boldsymbol{h} \\ & + (\Delta\sigma_3 + 2r^{-2}(\sigma_2 - \sigma_3)\boldsymbol{h}' \otimes \boldsymbol{h}' + (\Delta\sigma_4 - r^{-2}\sigma_4) \\ & (\boldsymbol{e}_1 \otimes \boldsymbol{h} + \boldsymbol{h} \otimes \boldsymbol{e}_1) \\ & + \alpha,_{zz}\, \boldsymbol{e}_1 \otimes \boldsymbol{e}_1 + \alpha,_{rr}\, \boldsymbol{h} \otimes \boldsymbol{h} + r^{-1}\alpha,_r\, \boldsymbol{h}' \otimes \boldsymbol{h}' + \alpha,_{zr} \\ & (\boldsymbol{e}_1 \otimes \boldsymbol{h} + \boldsymbol{h} \otimes \boldsymbol{e}_1). \end{aligned}$$

With this, we see that the four scalar consequences of the compatibility condition (2.68) are:

$$\begin{aligned} \Delta\sigma_1 + \alpha,_{zz} &= 0, \\ \Delta\sigma_2 - 2r^{-2}(\sigma_2 - \sigma_3) + \alpha,_{rr} &= 0, \\ \Delta\sigma_3 + 2r^{-2}(\sigma_2 - \sigma_3) + r^{-1}\alpha,_r &= 0, \\ \Delta\sigma_4 - r^{-2}\sigma_4 + \alpha,_{zr} &= 0 \end{aligned} \tag{5.31}$$

(cf. formulas (g), on p. 346 of [11]).

An inspection of system (5.31) leads to the following observations:

- on adding the first three equations, and on remembering (5.23), it is easy to check that $\Delta(\mathrm{tr}\,\mathbf{S}) = 0$; thus, in virtue of (5.23) itself,

$$-\alpha,_{zz} = \alpha,_{rr} + r^{-1}\alpha,_r\,; \tag{5.32}$$

- with the use of (5.32), summation of the second and third equations yields:

$$\Delta(\sigma_2 + \sigma_3) - \alpha,_{zz} = 0; \tag{5.33}$$

- on combining the fourth equation with (5.27), and on using $(5.30)_2$, we find, after some manipulations,

$$(\sigma_2 - (r\sigma_3),_r + \nu r\alpha,_r),_z = 0$$

  (cf. (5.19)).

In the light of these observations, we propose the following sequential procedure to solve (5.31):

1. to determine, a multiplicative constant apart, an appropriate solution $\alpha = \widehat{\alpha}(z, r)$ of the Laplace equation;
2. to integrate $(5.31)_1$ for σ_1:

$$\sigma_1 = -\Delta^{-1}[\alpha,_{zz}] + c_1\alpha, \tag{5.34}$$

   in a form where Δ^{-1} denotes the integral operator that formally inverts the laplacian, and where c_1 is a constant to be determined;
3. to integrate $(5.31)_4$ for σ_4;
4. to determine the fields $\widehat{\sigma}_2$ and $\widehat{\sigma}_3$, by solving the system of $(5.31)_2$ and (5.19):

$$\begin{aligned}\sigma_2 + \sigma_3 &= \Delta^{-1}[\alpha,_{zz}] + c_2\alpha,\\ \sigma_2 - (r\sigma_3),_r &+ \nu r\alpha,_r = 0,\end{aligned} \tag{5.35}$$

   where the constant c_2 is such that[8]

$$c_1 + c_2 = 1 + \nu. \tag{5.36}$$

We shall take these steps in the next two sections.

Remark 5.6 (cf. Remark 2 in [6].) For an alternative procedure, note the following consequence of (5.13) and (5.14):

$$\nabla(r\sigma_4) = -((r\sigma_2),_r - \sigma_3)\mathbf{e}_1 - r\sigma_1,_z\,\mathbf{h}. \tag{5.37}$$

[8] This condition is arrived at by adding (5.34) and $(5.35)_1$ and by taking into account of (5.30).

Were Step 4 taken right after Steps 1 and 2, the field $\widehat{\sigma}_4$ could be determined by integrating Eq. (5.37) along any regular curve $\mathcal{C}$, arc-length parameterized, with tangent $\boldsymbol{t}$, and going from a fixed point x_0 to the variable point x:

$$(r\sigma_4)|_{x_0}^{x} = -\int_{\mathcal{C}} ((r\sigma_2),_r - \sigma_3)\boldsymbol{e}_1 + r\sigma_1,_z \boldsymbol{h}) \cdot \boldsymbol{t}, ds. \tag{5.38}$$

Remarkably, given the extreme points, the choice of the joining curve is irrelevant, due to a well known result in the theory of differential forms: for R a *star-shaped* open region, and for $\boldsymbol{\omega} := \omega_1\boldsymbol{e}_1 + \omega_2\boldsymbol{e}_2 + \omega_3\boldsymbol{e}_3$ a vector field of class $C^1(R)$, the differential form $\omega = \omega_1 dx_1 + \omega_2 dx_2 + \omega_3 dx_3$ is exact if and only if $\operatorname{curl}\boldsymbol{\omega} = \boldsymbol{0}$ in R.[9] In our case, the differential form and associated vector field under scrutiny are:

$$\omega = \big((r\sigma_2),_r - \sigma_3\big)\, dz + r\sigma_1,_z\, dr, \quad \boldsymbol{\omega} = \big((r\sigma_2),_r - \sigma_3\big)\boldsymbol{e}_1 + r\sigma_1,_z\, \boldsymbol{h};$$

consequently,

$$\operatorname{curl}\boldsymbol{\omega} = \big(\sigma_1,_{zz} - ((r\sigma_2),_r - \sigma_3),_r\big)\boldsymbol{h}'.$$

Now, by differentiating (5.13) with respect to z and (5.14) with respect to r, and by subtracting the resulting relations, we obtain:

$$\sigma_1,_{zz} - ((r\sigma_2),_r - \sigma_3),_r = 0;$$

this result allows us to conclude that $\operatorname{curl}\boldsymbol{\omega} = \boldsymbol{0}$, and hence that the differential form ω is exact.

5.3.4 The Trace of the Stress Field

To determine the stress field solving the Boussinesq Problem, Step 1 consists in choosing (to within a constant) its trace, i.e., a harmonic field $\alpha = \widehat{\alpha}(z, r)$ defined over the half-space $\mathcal{HS}^+$.

We begin by observing that the four functions of the cylindrical coordinates (z, r), involved in the representation (5.12) of the stress field, can be substituted by an equal number of functions of the spherical coordinates (ρ, ϑ):

$$\widehat{\sigma}_i(z, r) = \widehat{\sigma}_i(\rho\cos\vartheta, \rho|\sin\vartheta|) =: \widetilde{\sigma}_i(\rho, \vartheta), \quad \widetilde{\sigma}_i(\rho, \vartheta) = \widetilde{\sigma}_i(\rho, -\vartheta), \ \forall\, \rho > 0. \tag{5.39}$$

Let us imagine to repeat the computation carried out in Sect. 5.2.1 to determine the equilibrium condition of the half-ball of radius ρ subject to the concentrated load shown in Fig. 5.3, this time for the stress field (5.12). It is convenient to write the outward normal to the hemisphere $\frac{1}{2}\mathcal{S}_\rho$ in the form:

---

[9] R is star-shaped if there is a point $p_0 \in R$ such that the line segment from p_0 to any point $p \in \partial R$ intersects ∂R only at p itself.

$$\widehat{\boldsymbol{n}}(\rho, \vartheta, \varphi) = \cos\vartheta \boldsymbol{e}_1 + |\sin\vartheta|\, \boldsymbol{h}(\varphi), \quad (\vartheta, \varphi) \in (0, +\pi/2) \times (0, 2\pi);$$

with this, the surface traction has the expression:

$$\boldsymbol{Sn} = (\cos\vartheta\, \sigma_1 + |\sin\vartheta|\sigma_4)\boldsymbol{e}_1 + (|\sin\vartheta|\sigma_2 + \cos\vartheta\, \sigma_4)\boldsymbol{h}.$$

In view of the boundary condition (5.17), we have that:

$$\begin{aligned} f\boldsymbol{e}_1 &= -\int_{\frac{1}{2}S_\rho} \boldsymbol{Sn}\, da = -\rho^2 \int_0^{+\pi/2} \Big(\int_0^{2\pi} \boldsymbol{Sn}\, d\varphi \Big) |\sin\vartheta|\, d\vartheta \\ &= -2\pi\rho^2 \Big(\int_0^{+\pi/2} (\cos\vartheta\, \widetilde{\sigma}_1(\rho, \vartheta) + |\sin\vartheta| \widetilde{\sigma}_4(\rho, \vartheta)) |\sin\vartheta| d\vartheta \Big) \boldsymbol{e}_1, \end{aligned} \tag{5.40}$$

whatever $\rho > 0$. Thus, for the right side to remain finite when ρ is chosen arbitrarily large, it is necessary that

$$\widetilde{\sigma}_1(\rho, \vartheta) = \rho^{-2}\widetilde{\tau}_1(\vartheta) + o(\rho^{-2});$$
$$\widetilde{\sigma}_4(\rho, \vartheta) = \rho^{-2}\widetilde{\tau}_4(\vartheta) + o(\rho^{-2}).$$

This result, when combined with the parity conditions formulated in (5.39), suggests the following *Ansatz*:

$$\widetilde{\sigma}_i(\rho, \vartheta) = \rho^{-2}\widetilde{\tau}_i(\vartheta), \quad \widetilde{\tau}_i(\vartheta) = \widetilde{\tau}_i(-\vartheta) \quad (i = 1, \ldots, 4). \tag{5.41}$$

As a consequence, we have for $\operatorname{tr}\boldsymbol{S}$ the following preliminary representation

$$\operatorname{tr}\boldsymbol{S} = \rho^{-2}\widetilde{\tau}(\vartheta), \quad \widetilde{\tau}(\vartheta) = \widetilde{\tau}(-\vartheta). \tag{5.42}$$

Now, we know from Sect. 5.2.2 that, for such a field to be harmonic, the function $\widetilde{\tau}$ must have the form (5.8). In conclusion, we are induced to choose:

$$\operatorname{tr}\boldsymbol{S} = \tau_0\, \rho^{-2} \cos\vartheta, \tag{5.43}$$

with τ_0 a constant to be determined, proportional to the applied load.

Remark 5.7 A further motivation for *Ansatz* (5.41) comes from the following considerations. Let us suppose to confine our anticipation to choosing only $\widetilde{\sigma}_1$, in the form $\widetilde{\sigma}_1(\rho, \vartheta) = \rho^{-2}\widetilde{\tau}_1(\vartheta)$. Then,

$$\Delta\sigma_1 = \rho^{-4}(2\tau_1 + \cot\vartheta\, \tau' + \tau_1'') \tag{5.44}$$

(cf. (5.7)). On the other hand, if $\alpha = \widehat{\alpha}(\rho, \vartheta)$, we find that

$$\alpha_{,zz} = \cos^2\vartheta\,\alpha_{,\rho\rho} + 2\rho^{-2}\sin\vartheta\cos\vartheta\,\alpha_{,\vartheta} + \rho^{-1}\sin^2\vartheta\,\alpha_{,\rho}$$
$$- 2\rho^{-1}\sin\vartheta\cos\vartheta\,\alpha_{,\rho\vartheta} + \rho^{-2}\sin^2\vartheta\,\alpha_{,\vartheta\vartheta}\,. \tag{5.45}$$

Then, in order to satisfy the first of compatibility relations (5.31) whatever the value of ρ, we are once again induced to choose for $\operatorname{tr}\boldsymbol{S}$ the preliminary representation (5.42). Having done this, to satisfy (5.33) identically we are induced to choose for the functions $\widetilde{\sigma}_2$ and $\widetilde{\sigma}_3$ the representations in (5.41). Finally, the necessity of the representation for $\widetilde{\sigma}_4$ follows from (5.38).

Remark 5.8 (This Remark is based on Appendix 2 of [6].) In cylindrical coordinates, and for α independent of φ, the Laplace equation reads:

$$\Delta\alpha = \alpha_{,zz} + \alpha_{,rr} + r^{-1}\alpha_{,r} = 0. \tag{5.46}$$

A method to solve (5.46) consists in looking for solutions, if any, of the form

$$\widehat{\alpha}(z,r) = \rho^a z^b r^c,$$

where exponents a, b, c are to be determined later. Given that

$$\rho_{,z} = \frac{z}{\rho}, \quad \rho_{,r} = \frac{r}{\rho},$$

we easily find that:

$$r^{-1}\alpha_{,r} = a\rho^{a-2}z^b r^c + c\rho^a z^b r^{c-2},$$
$$\alpha, rr = a(a-2)\rho^{a-4}z^b r^{c+2} + a(2c+1)\rho^{a-2}z^b r^c + c(c-1)\rho^a z^b r^{c-2},$$
$$\alpha, zz = a(a-2)\rho^{a-4}z^{b+2}r^c + a(2b+1)\rho^{a-2}z^b r^c + b(b-1)\rho^a z^{b-2}r^{c-2},$$

whence

$$\Delta\alpha = \rho^{a-2}z^b r^c\Big(a(a+2b+2c+1) + b(b-1)\rho^2 z^{-2} + c^2\rho^2 r^{-2}\Big).$$

Thus, for the field $\widehat{\alpha}$ to be harmonic in $\mathcal{HS}^+$, it is necessary that

$$a(a+2b+2c+1) + b(b-1)\rho^2 z^{-2} + c^2\rho^2 r^{-2} = 0, \quad \forall\, z, r > 0.$$

This condition is equivalent to the following system of algebraic conditions on exponents a, b, c:

$$c = 0, \quad b(b-1) = 0, \quad a(a+2b+2c+1) = 0.$$

Then,
(i) if $b = 0$, either $a = 0$ or $a = -1$; in the first instance, we obtain the solution:

$$\widehat{\alpha}_1(z,r) = \text{cost.},$$

in the second, the solution:

$$\widehat{\alpha}_2(z,r) = \rho^{-1};$$

(ii) if $b = 1$, then either $a = 0$ or $a = -3$; in the first instance, we obtain the solution:

$$\widehat{\alpha}_3(z,r) = z;$$

in the second, the solution:

$$\widehat{\alpha}_4(z,r) = \rho^{-3} z.$$

In conclusion, the desired field must have the following form:

$$\widehat{a}(z,r) = \alpha_0 \frac{z}{\rho^3} + \alpha_1 \frac{1}{\rho} + \alpha_2 z + \alpha_3. \qquad (5.47)$$

The reader should notice that

$$\alpha_{3,z} = \alpha_1; \quad \alpha_{2,z} = -\alpha_4.$$

These results help to exemplify a general property: if $\widehat{\alpha}(z,r)$ is harmonic (i.e., in this context, if it satisfies Laplace equation in cylindrical coordinates), then $\widehat{\alpha}_{,z}$ is harmonic as well.

The general representation (5.47) is parameterized by four constants, whose choice is made on the basis of the conditions prevailing at the boundary of the region of interest. In our case, both constants α_2 and α_3 must be taken null, so as to comply with the physical palusibility requirement that the stress field—and hence its trace—vanish at infinity. The same must be done for α_1, this time on the basis of an application of a result due to Antonio Signorini (1888–1963), that we now recall in a version appropriate to our present context (see [8, Sect. 18]).

Signorini's Lemma. Let $\boldsymbol{S}$ be a stress field that balances the distance and contact forces $\boldsymbol{d}$ and $\boldsymbol{c}$ acting on a domain R with smooth boundary ∂R:

$$\operatorname{div} \boldsymbol{S} + \boldsymbol{d} = \boldsymbol{0} \ \text{ in } R, \qquad \boldsymbol{S}\boldsymbol{n} = \boldsymbol{c} \ \text{ on } \partial R.$$

Moreover, let $\boldsymbol{w}$ be a smooth vector field on $R \cup \partial R$. Then,

$$\int_R (\nabla \boldsymbol{w})\boldsymbol{S}\, dv = \int_R \boldsymbol{w} \otimes \boldsymbol{d}\, dv + \int_{\partial R} \boldsymbol{w} \otimes \boldsymbol{c}\, da. \qquad (5.48)$$

We specialize (5.48) for $R \equiv \mathcal{B}_\rho$, a half-ball of radius ρ centered at the point of application of the load, $\boldsymbol{d} \equiv \boldsymbol{0}$, and $\boldsymbol{w} = \mathbf{x}$. Taking the trace of the resulting identity, we obtain:

$$\int_{\mathcal{B}_\rho} \operatorname{tr} \boldsymbol{S}\, dv = \int_{\partial\mathcal{B}_\rho} \rho \boldsymbol{n} \cdot \boldsymbol{S}\boldsymbol{n}\, da.$$

Now, the right side of this integral identity is of order $O(\rho)$, because we have from (5.40) that

$$\int_{\partial\mathcal{B}_\rho} \boldsymbol{S}\boldsymbol{n}\, da = O(1).$$

As to the left side, we have that

$$\int_{\mathcal{B}_\rho} \operatorname{tr} \boldsymbol{S}\, dv = 2\pi(1+\nu) \int_0^\rho \int_0^\pi (\alpha_0 \cos\vartheta + s\alpha_1)\, d\vartheta\, ds;$$

for it to be of order $O(\rho)$ as well, α_1 must be set equal to 0. We conclude that (5.47) reduces to

$$\alpha = \widehat{a}(z,r) = \alpha_0 \frac{z}{\rho^3}; \tag{5.49}$$

in spherical coordinates, representation (5.49) translates into

$$\alpha = \widetilde{\alpha}(\rho,\vartheta) = \alpha_0\, \rho^{-2} \cos\vartheta. \tag{5.50}$$

(cf. (5.43)).

Remark 5.9 A different way to look for harmonic functions with cylindrical symmetry consists in using a preliminary representation with separate variables:

$$\widehat{\alpha}(z,r) = Z(z)R(r).$$

In this case, the unknown functions Z and R have to satisfy the differential relation:

$$\frac{R''}{R} + r^{-1}\frac{R'}{R} = -\frac{Z''}{Z} = -\kappa^2, \quad \forall\, z, r > 0,$$

which is equivalent to the following ODEs:

$$R'' + r^{-1}R' + \kappa^2 R = 0, \quad Z'' - \kappa^2 Z = 0.$$

The first is Bessel equation, whose solutions are the zero-order Bessel and Neumann functions $J_0(\kappa r)$ and $N(\kappa r) = (2/\pi)J_0(\kappa r)\ln r$; the second has the solutions $\exp(\pm\kappa z)$.

5.4 The Boussinesq Stress Field

On arriving at the alternative representations (5.47) and (5.50), the first of the four steps in the sequential procedure laid down when closing Sect. 5.3.3 was completed. We now take the remaining steps.

5.4.1 Step 2

This step consists in solving for σ_1 Eq. (5.31)$_1$, that we here reproduce:

$$\Delta\sigma_1 + \alpha,_{zz} = 0.$$

With the use of representations (5.41)$_1$ and (5.50) for σ_1 and α, and on accounting for both (5.44) and (5.45), Eq. (5.31)$_1$ takes the following form:

$$\tau_1'' + \cot\vartheta\,\tau_1' + 2\tau_1 + 3\alpha_0\cos\vartheta(2\cos^2\vartheta - 3\sin^2\vartheta) = 0,$$

a inhomogeneous second-order ODE with non constant coefficients. A particular solution can be obtained, for example, with the *variation-of-constants method*, given that fundamental solutions and wronskian of the associated homogeneous equation are known to us.[10] We find:

$$\widetilde{\tau}_1(\vartheta) = \frac{3}{2}\alpha_0\cos^3\vartheta.$$

By addition of a multiple of $\cos\vartheta$, that is, of an even solution of the homogeneous equation, we end up with:

$$\widetilde{\sigma}_1(\rho,\vartheta) = \rho^{-2}\Big(\frac{3}{2}\alpha_0\cos^3\vartheta + \beta_0\cos\vartheta\Big). \tag{5.51}$$

5.4.2 Step 3

The representation of function $\widetilde{\tau}_4$ is determined by the ODE:

$$\tau_4'' + \cot\vartheta\,\tau_4' + (1-\cot^2\vartheta)\tau_4 + 3\alpha_0|\sin\vartheta|(4\cos^2\vartheta - \sin^2\vartheta) = 0;$$

for $\vartheta \in (-\pi/2, 0) \cup (0, +\pi/2)$, a particular solution is:

$$\widetilde{\tau}_4(\vartheta) = \frac{3}{2}\alpha_0\cos^2\vartheta\,|\sin\vartheta| \tag{5.52}$$

[10] Revert to the footnote in Sect. 5.2.2.

(the proof of the pudding is in the eating), while the even solution of the associated homogeneous equation is $|\sin\vartheta|$. Thus,

$$\widetilde{\sigma}_4(\rho,\vartheta)=\rho^{-2}\left(\frac{3}{2}\alpha_0\cos^2\vartheta|\sin\vartheta|+\gamma_0|\sin\vartheta|\right). \tag{5.53}$$

We are now in a position to show that, for equilibrium, both constants β_0 in (5.51) and γ_0 on (5.53) must be null. Firstly, we observe that:

$$\widehat{\sigma}_1(z,r)=\frac{3}{2}\alpha_0\frac{z^3}{(z^2+r^2)^{\frac{5}{2}}}+\beta_0\frac{z}{(z^2+r^2)^{\frac{3}{2}}},$$
$$\widehat{\sigma}_4(z,r)=\frac{3}{2}\alpha_0\frac{z^2r}{(z^2+r^2)^{\frac{5}{2}}}+\gamma_0\frac{r}{(z^2+r^2)^{\frac{3}{2}}}.$$

Consequently, the balance equation $(5.13)_1$ is satisfied (if and) only if

$$\beta_0=\gamma_0.$$

Secondly, we observe that $\widehat{\sigma}_4(0,r)$, and hence γ_0, must be null, in order to satisfy the second of conditions (5.16) (recall that (5.16) expresses the requirement that the traction-vector field be null over the load-free boundary plane). We conclude that $\beta_0=\gamma_0=0$.

We can now take up again (5.40), and deduce from it that[11]

$$\alpha_0=-\frac{f}{\pi}.$$

In conclusion,

$$\widetilde{\alpha}(\rho,\vartheta)=\frac{f}{\pi}\rho^{-2}\cos\vartheta, \tag{5.54}$$

and then, on recalling (5.30),

$$\mathrm{tr}\,\mathbf{S}=-\frac{f}{\pi}(1+\nu)\rho^{-2}\cos\vartheta.$$

Moreover, (5.51) becomes:

$$\widetilde{\sigma}_1(\rho,\vartheta)=-\frac{3f}{2\pi}\rho^{-2}\cos^3\vartheta. \tag{5.55}$$

[11] On taking both (5.51) and (5.52) into account, (5.40) becomes:

$$f=-2\pi\int_0^{+\pi/2}(\cos\vartheta\,\widetilde{\tau}_1(\vartheta)+|\sin\vartheta|\widetilde{\tau}_4(\vartheta))|\sin\vartheta|d\vartheta=-\pi\alpha_0.$$

5.4.3 Step 4

It follows from (5.34), $(5.35)_1$, and (5.36), that

$$\sigma_2 + \sigma_3 = -\sigma_1 + (1+\nu)\alpha;$$

and, on reverting to cylindrical coordinates, (5.55) and (5.54) yield:

$$\widehat{\sigma}_1(z,r) = -\frac{3}{2}\frac{f}{\pi}\frac{z^3}{\rho^5}, \quad \widehat{\alpha}(z,r) = -\frac{f}{\pi}\frac{z}{\rho^3}.$$

Thus, we have that

$$\sigma_2 = -\sigma_3 + \frac{3}{2}\frac{f}{\pi}\frac{z^3}{\rho^5} - \frac{f}{\pi}(1+\nu)\frac{z}{\rho^3}. \tag{5.56}$$

Substituting this expression for σ_2 into the second of (5.35), for each fixed value of z we arrive at the following ODE for σ_3 as a function of r:

$$\sigma_3 + (r\sigma_3)_{,r} = \frac{3}{2}\frac{f}{\pi}\frac{z^3}{\rho^5} - \frac{f}{\pi}(1+\nu)\frac{z}{\rho^3} - \frac{3f}{\pi}\nu\frac{rz}{\rho^5}, \tag{5.57}$$

whose solution is:

$$\widehat{\sigma}_3(z,r) = \frac{f}{2\pi}(1-2\nu)\frac{z(z^2+2r^2)}{r^2\rho^3} + \frac{g(z)}{r^2}, \tag{5.58}$$

where g is an arbitrary function; finally, by substituting (5.58) into (5.56) we have:

$$\widehat{\sigma}_2(z,r) = \frac{3}{2}\frac{f}{\pi}\frac{z^3}{\rho^5} - \frac{f}{\pi}(1+\nu)\frac{z}{\rho^3} - (1-2\nu)\frac{f}{2\pi}\frac{z(z^2+2r^2)}{r^2\rho^3} - \frac{g(z)}{r^2}. \tag{5.59}$$

Remark 5.10 We find it important to point out a difference between the Flamant Problem, where the stress field can be completely determined irrespectively of the constitutive response, and the Boussinesq Problem, where the stress field is determined to within a bit of information, the function g in (5.58) and (5.59), whose determination requires the imposition of a *kinematic symmetry condition* that, as we shall see in the next section, to be effective must be formulated with the use of the inverse constitutive law $(2.45)_2$. The condition in question is:

$$\lim_{r\to 0^+} u_r = 0, \tag{5.60}$$

requiring that, on whatever horizontal plane, the radial displacement component vanishes with the distance from the origin. We shall see that it implies that

$$g(z) \equiv -\frac{f}{2\pi}(1-2\nu). \tag{5.61}$$

In conclusion, on taking (5.61) into account, the solution of system (5.31) is:

$$\begin{aligned}
(2\pi f^{-1})\,\widehat{\sigma}_1^B(z,r) &= -3\frac{z^3}{(z^2+r^2)^{5/2}},\\
(2\pi f^{-1})\,\widehat{\sigma}_2^B(z,r) &= (1-2\nu)\frac{1}{(z^2+r^2)^{1/2}((z^2+r^2)^{1/2}+z)}-3\frac{zr^2}{(z^2+r^2)^{5/2}},\\
(2\pi f^{-1})\,\widehat{\sigma}_3^B(z,r) &= (1-2\nu)\Big(\frac{z}{(z^2+r^2)^{3/2}}-\frac{1}{(z^2+r^2)^{1/2}((z^2+r^2)^{1/2}+z)}\Big),\\
(2\pi f^{-1})\,\widehat{\sigma}_4^B(z,r) &= -3\frac{z^2 r}{(z^2+r^2)^{5/2}}
\end{aligned} \tag{5.62}$$

(cf. [9, p. 565], [11, p. 364, eq. (201)]), so that the related stress field is

$$\boldsymbol{S}^B(z,r,\varphi) = \sigma_1^B \boldsymbol{e}_1\otimes\boldsymbol{e}_1 + \sigma_2^B \boldsymbol{h}\otimes\boldsymbol{h} + \sigma_3^B \boldsymbol{h}'\otimes\boldsymbol{h}' + \sigma_4^B(\boldsymbol{e}_1\otimes\boldsymbol{h}+\boldsymbol{h}\otimes\boldsymbol{e}_1), \tag{5.63}$$

where

$$\sigma_1^B = S_{zz}^B,\quad \sigma_2^B = S_{rr}^B,\quad \sigma_3^B = S_{\varphi\varphi}^B,\quad \sigma_4^B = S_{r\varphi}^B = S_{\varphi r}^B. \tag{5.64}$$

Alternatively, with the use of (5.39), we can write the solution of system (5.31) as

$$\begin{aligned}
(2\pi f^{-1})\rho^2\widetilde{\sigma}_1^B(\rho,\vartheta) &= -3\cos^3\vartheta,\\
(2\pi f^{-1})\rho^2\widetilde{\sigma}_2^B(\rho,\vartheta) &= (1-2\nu)\frac{1}{1+\cos\vartheta}-3\sin^2\vartheta\cos\vartheta,\\
(2\pi f^{-1})\rho^2\widetilde{\sigma}_3^B(\rho,\vartheta) &= (1-2\nu)\Big(\cos\vartheta-\frac{1}{1+\cos\vartheta}\Big),\\
(2\pi f^{-1})\rho^2\widetilde{\sigma}_4^B(\rho,\vartheta) &= -3|\sin\vartheta|\cos^2\vartheta.
\end{aligned}$$

Remark 5.11 We see at a glance that boundary conditions (5.16) are satisfied.

5.5 Digression

5.5.1 Concentrated Contact Interactions

One may ask whether concentrated contact interactions arise in Boussinesq's problem. Guided by the experience gained with the Flamant Problem, where the overall symmetries of the stress field are very similar, we check the translational equilibrium in the $\boldsymbol{e}_2$-direction of a half of the hemisphere $\frac{1}{2}\mathcal{S}_\rho$ (Fig. 5.6). Our understanding is that, in case a concentrated contact interaction $\boldsymbol{f}_c = f_c\boldsymbol{e}_2$ were needed to guarantee that equilibrium, it would be found at the origin.

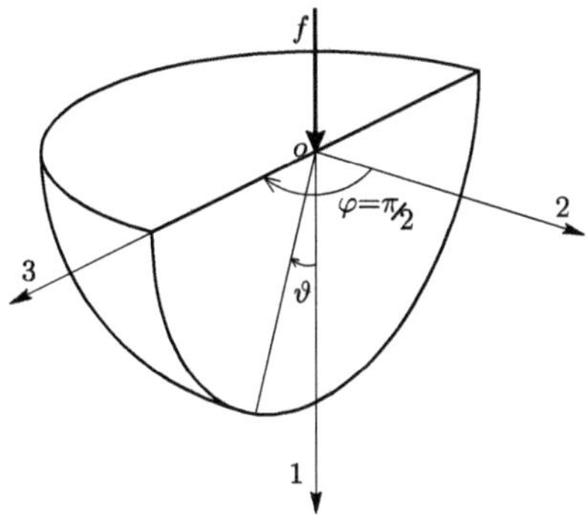

Fig. 5.6 Equilibrium of half of a hemispherical part

Now, in the plane whose outer normal is $\boldsymbol{e}_2$, the $\boldsymbol{e}_2$-component of the traction vector is:

$$\boldsymbol{S}^B\boldsymbol{e}_2 \cdot \boldsymbol{e}_2 = \big(\sigma_2 \cos^2\varphi + \sigma_3 \sin^2\varphi\big)|_{\varphi=\pm\pi/2} = \sigma_3;$$

its resultant is null:

$$\int_0^\rho \int_{-\pi/2}^{+\pi/2} \sigma_3(t,\vartheta) t\, d\vartheta dt = \int_0^\rho \int_{-\pi/2}^{+\pi/2} t^{-1} \left(\cos\vartheta - \frac{1}{1+\cos\vartheta}\right) d\vartheta dt = 0,$$

whereas the $\boldsymbol{e}_2$-component of the resultant traction on the spherical part of the boundary is not:

$$\begin{aligned}
&\left(\int_{\pi/2}^{3\pi/2} \int_0^{\pi/2} \widetilde{\boldsymbol{S}}^B(\rho,\vartheta)\boldsymbol{n}(\vartheta,\varphi)\rho^2 |\sin\vartheta|\, d\vartheta d\varphi\right) \cdot \boldsymbol{e}_2 \\
&= \left(\int_{\pi/2}^{3\pi/2} \int_0^{\pi/2} (|\sin\vartheta|\sigma_2 + \cos\vartheta\,\sigma_4)\rho^2 |\sin\vartheta| \boldsymbol{h}(\varphi)\right) \cdot \boldsymbol{e}_2 \\
&= -2\int_0^{\pi/2} (|\sin\vartheta|\sigma_2 + \cos\vartheta\,\sigma_4)\rho^2 |\sin\vartheta| = \frac{f}{2\pi}\big(4 - \pi + 2\nu(\pi - 2)\big).
\end{aligned}$$

We conclude that, in order to assure the equilibrium of the body part under examination when isolated from the rest by an Euler cut, a concentrated interaction arises at the point where the stress field is singular, namely,

$$\boldsymbol{f}_c = -\frac{f}{2\pi}\big(4 - \pi + 2\nu(\pi - 2)\big)\boldsymbol{e}_2.$$

Interestingly, unlike with Flamant's and Cerruti's problems (see, respectively, Chap. 4, Sect. 4.4.2 and Chap. 8, Remark 8.1), both intensity and orientation of such a concentrated interaction depend on the material, through the value of its Poisson ratio; in particular, it points in the opposite direction of the $\boldsymbol{e}_2$-axis for $\nu \in (-(4-\pi)/2(\pi-2), 1/2)$.

5.5.2 How to Find Flamant's Stress Field Having Found Boussinesq's

Knowledge of the elastic state that solves Boussinesq's problem makes the solution of Flamant's problem an easy exercise.[12] The procedure is identical to the one sketched in our introductory Chap. 1 for a beam subject to axial loads: Boussinesq's displacement or stress fields are given the role of Green functions (Boussinesq's 'potentials') for the corresponding Flamant's fields.

Suppose, for example, that we are asked to find a component of Flamant's stress tensor $\boldsymbol{S}^F$ – S_{11}^F, say – starting from the knowledge of the same component of $\boldsymbol{S}^B$, that for convenience we represent provisionally in Cartesian coordinates:

$$S_{11}^B(x_1, x_2, x_3) = -\frac{3f}{2\pi}\frac{x_1^3}{(x_1^2+x_2^2+x_3^2)^{5/2}}\,.$$

Just as we did in Remark 1.5, we think of a unit concentrated load applied at a point of the x_3-axis of coordinates $(0, 0, \zeta)$; consequently,

$$S_{11}^B(x_1, x_2, x_3; \xi) = -\frac{3}{2\pi}\frac{x_1^3}{(x_1^2+x_2^2+(x_3-\xi)^2)^{5/2}}\,.$$

The effect of the Flamant line loading is reproduced by superposition, at the expenses of computing a Cauchy's integral over the x_3-axis:

$$\begin{aligned}S_{11}^F &= \int_{-\infty}^{+\infty} S_{11}^B(x_1, x_2, x_3; \xi)d\xi = -\frac{f}{2\pi}\frac{x_1^3}{(x_1^2+x_2^2)^2}\\ &\times \lim_{n\to\infty}\left(\frac{3(x_1^2+x_2^2)+2(x_3-n)^2}{(x_1^2+x_2^2+(x_3-n)^2)}(x_3-n)+\frac{3(x_1^2+x_2^2)+2(x_3+n)^2}{(x_1^2+x_2^2+(x_3+n)^2)}(x_3+n)\right)\\ &= -\frac{2f}{\pi}\frac{x_1^3}{(x_1^2+x_2^2)^2} = -\frac{2f}{\pi}\rho^{-1}\cos^3\vartheta.\end{aligned}$$

Needless to say, the result is the one expected.

5.6 The Boussinesq Strain and Displacement Fields

On making use of the inverse constitutive equation $(2.45)_2$ and relation (2.43) between the technical moduli, we have that:

[12] In [7], Flamant himself recognizes his debts to Boussinesq.

$$
\begin{aligned}
E_{zz} &= \frac{1}{2G}\Big(S_{zz} - \frac{\nu}{1+\nu}(S_{zz}+S_{rr}+S_{\varphi\varphi})\Big),\\
E_{rr} &= \frac{1}{2G}\Big(S_{rr} - \frac{\nu}{1+\nu}(S_{zz}+S_{rr}+S_{\varphi\varphi})\Big),\\
E_{\varphi\varphi} &= \frac{1}{2G}\Big(S_{\varphi\varphi} - \frac{\nu}{1+\nu}(S_{zz}+S_{rr}+S_{\varphi\varphi})\Big),\\
E_{zr} &= \frac{1}{2G}S_{zr}, \quad E_{z\varphi} = E_{\varphi r} = 0.
\end{aligned}
$$

Combining these relations with (5.64) and (5.62), we find:

$$
\begin{aligned}
E_{zz} &= \frac{f}{4\pi G}\Big(-3\frac{z^3}{\rho^5} + 2\nu\frac{z}{\rho^3}\Big),\\
E_{rr} &= \frac{f}{4\pi G}\Big(4(1-\nu)\frac{zr^2}{\rho^5} + (1-2\nu)\frac{z^5}{r^2\rho^5} + 2(1-3\nu)\frac{z^3}{\rho^5}\Big) - \frac{1}{2G}\frac{g(z)}{r^2},\\
E_{\varphi\varphi} &= \frac{f}{4\pi G}\Big(2(1-\nu)\frac{z}{\rho^3} + (1-2\nu)\frac{z^3}{r^2\rho^3}\Big) + \frac{1}{2G}\frac{g(z)}{r^2},\\
E_{zr} &= -\frac{3f}{4\pi G}\frac{z^2 r}{\rho^5},
\end{aligned}
\tag{5.65}
$$

where, we recall, $\rho^2 = z^2 + r^2$.

Now, we know from (5.11) that the displacement field we seek must have the form:

$$
\boldsymbol{u} = \widehat{\boldsymbol{u}}(z, r, \varphi) = u_z(z, r)\boldsymbol{e}_1 + u_r(z, r)\boldsymbol{h}(\varphi),
$$

which implies that

$$
\nabla \boldsymbol{u} = u_{z,z}\,\boldsymbol{e}_1 \otimes \boldsymbol{e}_1 + u_{r,z}\,\boldsymbol{h} \otimes \boldsymbol{e}_1 + r^{-1}u_r\boldsymbol{h}' \otimes \boldsymbol{h}' + u_{z,r}\,\boldsymbol{e}_1 \otimes \boldsymbol{h} + u_{r,r}\,\boldsymbol{h} \otimes \boldsymbol{h}.
$$

With this, given the definition $(2.2)_2$ of $\boldsymbol{E}$, we arrive at:

$$
E_{zz} = u_{z,z}, \quad E_{rr} = u_{r,r}, \quad E_{\varphi\varphi} = r^{-1}u_r, \quad E_{zr} = \frac{1}{2}\Big(u_{z,r} + u_{r,z}\Big). \tag{5.66}
$$

An expression for u_r is found *via* the algebraic combination of $(5.65)_3$ and $(5.66)_3$:

$$
u_r = \frac{f}{4\pi G}\Big(2(1-\nu)\frac{zr}{\rho^3} + (1-2\nu)\frac{z^3}{r\rho^3}\Big) + \frac{1}{2G}\frac{g(z)}{r}; \tag{5.67}
$$

from this, we see that, to satisfy (5.60), g must be chosen as specified in (5.61). Hence,

$$
\widehat{u}_r^B(z, r) = \frac{f}{4\pi G}\Big(\frac{zr}{\rho^3} - (1-2\nu)\frac{r}{\rho(\rho+z)}\Big). \tag{5.68}
$$

Next, given that $(5.66)_4$ implies that

$$u_z = \int (2E_{rz} - u_{r,z})dz,$$

we compute $u_{r,z}$ from (5.68) and use of $(5.65)_4$, so as to obtain:

$$u_z = \frac{f}{4\pi G}\Big(\frac{z^2}{\rho^3} + 2(1-\nu)\frac{1}{\rho}\Big) + h(z).$$

A computation of $u_{z,z}$ based on this expression is consistent with:

$$u_{z,z} = \frac{f}{4\pi G}\Big(-3\frac{z^3}{\rho^5} + 2\nu\frac{z}{\rho^3}\Big)$$

– that is, with the combination of $(5.66)_1$ and $(5.65)_1$ – only if $h(z) \equiv 0$. We conclude that

$$\widehat{u}_z^B(z,r) = \frac{f}{4\pi G}\Big(\frac{z^2}{\rho^3} + 2(1-\nu)\frac{1}{\rho}\Big). \tag{5.69}$$

Remark 5.12 Formulas (5.68) and (5.69) correspond, respectively, to formulas (203) and (204) at p. 365 of [11]. They can be rewritten in a form:

$$\begin{aligned} u_r^B &= \frac{f}{4\pi G}\rho^{-1}|\sin\vartheta|\Big(\cos\vartheta - (1-2\nu)\frac{1}{1+\cos\vartheta}\Big),\\ u_z^B &= \frac{f}{4\pi G}\rho^{-1}\Big(\cos^2\vartheta + 2(1-\nu)\Big), \end{aligned} \tag{5.70}$$

that highlights the behaviour of the displacement field near the origin, where a *first-order singularity* occurs, and at infinity, where the displacement field vanishes. Similarly, writing (5.65) in the form:

$$\begin{aligned} \rho^2 E_{zz}^B &= \frac{f}{4\pi G}(-3\cos^3\vartheta + 2\nu\cos\vartheta),\\ \rho^2 E_{rr}^B &= \frac{f}{4\pi G}\Big((1-2\nu)\frac{1}{1+\cos\vartheta} - 3\cos\vartheta\sin^2\vartheta + 2\nu\cos\vartheta\Big),\\ \rho^2 E_{\varphi\varphi}^B &= \frac{f}{4\pi G}\Big(\cos\vartheta - (1-2\nu)\frac{1}{1+\cos\vartheta}\Big),\\ \rho^2 E_{zr}^B &= \frac{f}{4\pi G}(-3|\sin\vartheta|\cos^2\vartheta), \end{aligned} \tag{5.71}$$

highlights both the order of singularity at the origin—the same as for the stress field—and the fact that the strain field is null at infinity.

Remark 5.13 An alternative expression for the radial displacement is:

$$u_r^B = \frac{f}{8\pi G}\rho^{-1}\left(|\sin 2\vartheta| - 2(1-2\nu)\tan\frac{\vartheta}{2}\right)$$
$$= \frac{f}{4\pi G}\rho^{-1}\left(\cos^2\vartheta + \cos\vartheta - (1-2\nu)\right)\tan\frac{\vartheta}{2}.$$

We see that the locus of points whose horizontal displacement is null is the cone about the x_1 axis of aperture

$$\vartheta = \arccos\left(\frac{1}{2}\big(-1+\sqrt{1+4(1-2\nu)}\big)\right);$$

the points outside this cone tend to approach the axis, those inside move away from it.

Remark 5.14 On the boundary plane of $\mathcal{HS}^+$, the Boussinesq displacement field has the following form:

$$\widehat{\boldsymbol{u}}^B(0, r, \varphi) = \frac{f}{4\pi G}r^{-1}\big(2(1-\nu)\boldsymbol{e}_1 - (1-2\nu)\widehat{\boldsymbol{h}}(\varphi)\big), \tag{5.72}$$

whence, in particular,

$$\lim_{r\to+\infty}\widehat{\boldsymbol{u}}^B(0, r, \varphi) = \boldsymbol{0}.$$

We read from (5.72), just as physical intuition suggests, that *the vertical displacement has the direction and orientation of the load*, while *the radial displacement is directed toward the origin*; and that, as pointed out in Remark 5.12, both components of the displacement vector have a first-order singularity in the origin. Since the component ratio does not depend on r, on each half-line through the origin the direction of the surface displacement is constant, while the absolute value goes as r^{-1}; the two half-lines become the hyperbola shown in Fig. 5.7.

Finally, if we integrate the displacement field along the boundary of a surface disk of radius r, centered at the origin, we find:

$$\int_0^{2\pi}\widehat{\boldsymbol{u}}(0, r, \varphi)\, r d\varphi = \frac{(1-\nu)f}{G}\boldsymbol{e}_1.$$

Hence, the average radial displacement is null, and the average vertical displacement has constant value.

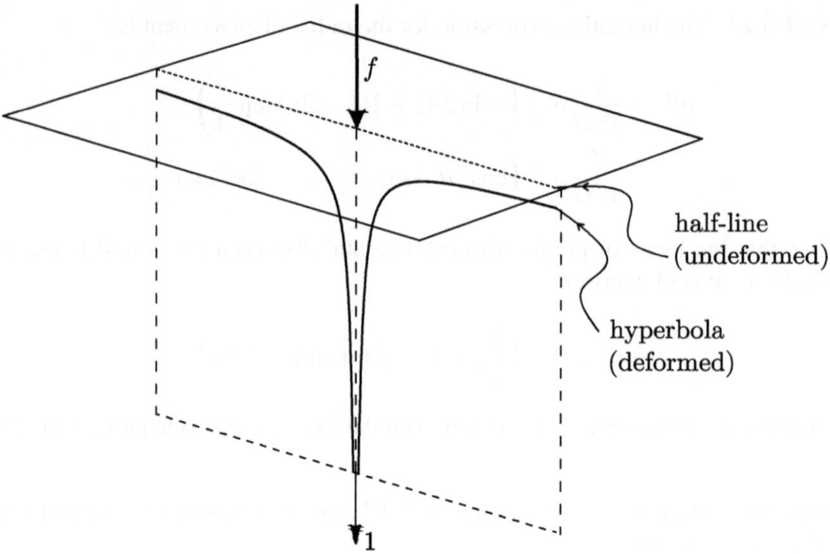

Fig. 5.7 The deformed shapes of two half-lines of the boundary plane through the point where the load is applied

5.7 Exploiting the Boussinesq Solution

In this section we discuss some simple problems, all of them of interest in applications, that can be solved explicitly with a more or less direct use of the solution to the Boussinesq Problem.

5.7.1 The Pressure Bulb

As in the case of the Flamant Problem (recall Remark 4.6), it is of interest for geotechnical applications to find the locus of points where the stress has a magnitude of given constant value, that is to say, the surface whose equation is $|\boldsymbol{S}^B|^2 = c^2$. A simple computation yields:

$$9\cos^6\vartheta + (1-2\nu)^2\left(\cos\vartheta - \frac{1}{1+\cos\vartheta}\right)^2 + 9\cos^4\vartheta\sin^2\vartheta$$
$$+\left(\frac{1-2\nu}{1+\cos\vartheta} - 3\cos\vartheta\sin^2\vartheta\right)^2 = c^2,$$

a revolution surface about the x_1-axis, whose trace on a typical plane through that axis is sketched in Fig. 5.8.

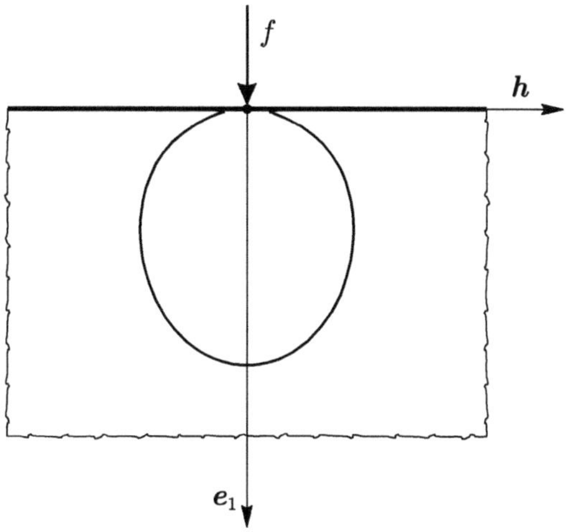

Fig. 5.8 The shape of the pressure bulb for the Boussinesq Problem

5.7.2 Displacements in a Half-Space Subject to a Diffused Load

Both this example and the one dealt with in the next subsection are taken from Section 124 of [11]. In both cases, the half-space $\mathcal{HS}^+$ is subject to a uniform diffused load p over a circle of radius r_0, centered at the origin. Our goal is to compute the vertical displacement of a typical point x of the region of plane $z = 0$ exterior to the load circle.

Let $r := |x - o|$, and let ψ_0 denote half the aperture of the angle formed by the tangents to the circle passing through x (Fig. 5.9). Moreover, let $s := |y - x|$ be the distance from x of a point y on the half-line through x forming an angle $\psi \in [-\psi_0, +\psi_0]$ with the half-line passing through x and o; it is not difficult to see that length of the chord the half-line of angle ψ intercepts with the circle is $2(r_0^2 - r^2 \sin^2 \psi)^{1/2}$. Now, we have from (5.72) that the vertical displacement induced at x by a vertical unit force applied at a distance r from it is:

$$u_1 = \frac{1-\nu}{2\pi G} r^{-1}; \tag{5.73}$$

therefore, the vertical displacement at x due to an infinitesimal force $p\,da = ps\,d\psi\,ds$ concentrated at y is:

$$d\,u_1 = \frac{1-\nu}{2\pi G} p\,d\psi\,ds.$$

Consequently, in view of the linearity of the Boussinesq Problem, the displacement induced by the whole applied load is:

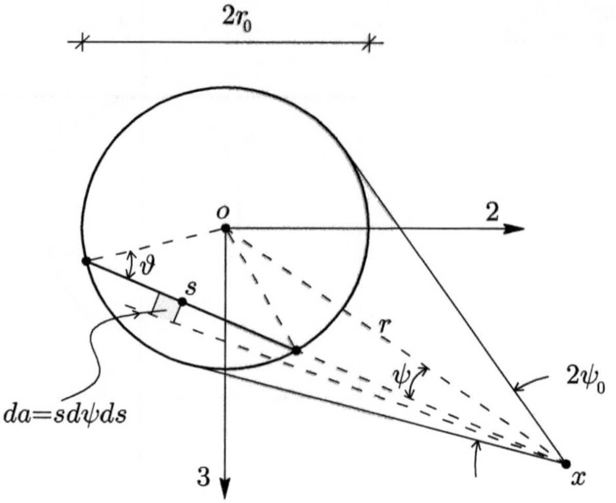

Fig. 5.9 The load circle on plane $x = 0$

$$u_1 = \frac{1-\nu}{2\pi G} p \int_{-\psi_0}^{+\psi_0} \int_0^{2\sqrt{r_0^2 - r^2 \sin^2\psi}} d\psi\, ds = \frac{2(1-\nu)}{\pi G} p \int_0^{+\psi_0} \sqrt{r_0^2 - r^2 \sin^2\psi}\, d\psi;$$

a change of variable suggested by the geometric relation:

$$r_0 \sin\vartheta = r \sin\psi, \quad \vartheta \in [0, \pi/2],$$

permits to lend this expression a form more convenient for computations:

$$u_1 = \frac{2(1-\nu)}{\pi G} pr \Big[\int_0^{\pi/2} \Big(1 - \frac{r_0^2}{r^2}\sin^2\vartheta\Big)^{1/2} d\vartheta - \Big(1 - \frac{r_0^2}{r^2}\Big) \int_0^{\pi/2} \Big(1 - \frac{r_0^2}{r^2}\sin^2\vartheta\Big)^{-1/2} d\vartheta \Big] \tag{5.74}$$

(cf. relation (206) at p. 367 of [11], which is obtained by an use of (2.43)).

For the vertical displacement of a boundary point of the load circle, formula (5.74) gives:

$$u_1(p, r_0) = \frac{2(1-\nu)}{\pi G} pr_0,$$

a value that one may wish to compare with

$$u_1(f, o) = (p(\pi r_0^2)) \frac{1-\nu}{2\pi G} r_0^{-1} = \frac{1-\nu}{2G} pr_0,$$

that is, according to (5.73), the displacement due to a force

$$f = p\left(\pi r_0^2\right)$$

concentrated at the origin: the ratio is $4/\pi$. These values can be compared with the value of the vertical displacement at the origin due a diffused circle load of intensity p, which turns out to be:

$$u_1 = \frac{1-\nu}{G} p r_0,$$

(see the cited section of [11]), that is, $\pi/2$ times the vertical displacement at the origin.

These results may serve to estimate the displacement at the bottom of a circular flexible foundation applying a reasonably uniform load to the underlying soil (think of an inflatable swimming pool, filled up with water, in the absence of a bunch of kids jumping up and down).

5.7.3 Displacements in a Half-Space Subject to a Diffused Load

To evaluate the stress field induced at a typical point $(z, 0, 0)$ of the vertical axis by the same diffused load as in the previous subsection, we consider the infinitesimal load portion over the ring between radius s and radius $s + ds$ (Fig. 5.10), whose effect is:

$$d\mathbf{S} = (f^{-1}\mathbf{S}^B(z, s, \varphi))ps\, d\varphi\, ds,$$

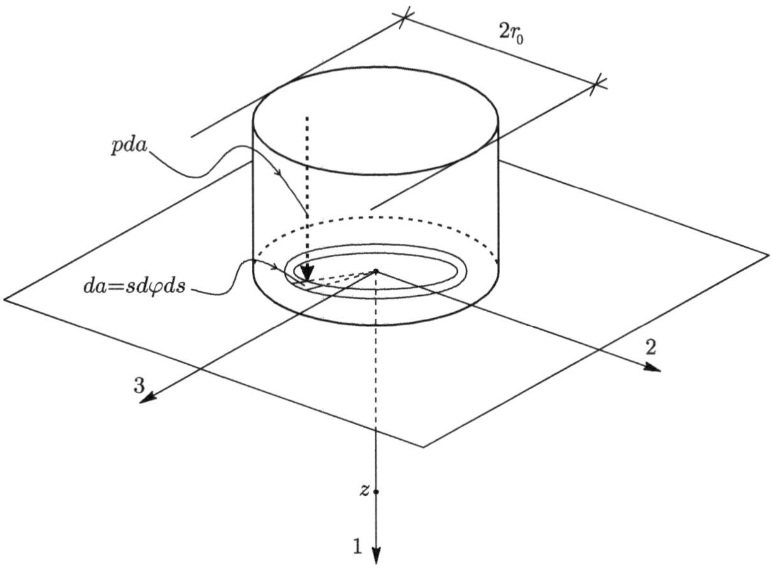

Fig. 5.10 A center-symmetric infinitesimal portion of a diffused vertical load

for the field $\boldsymbol{S}^B$ specified by (5.62) and (5.63). On recalling that

$$\int_0^{2\pi} \boldsymbol{h}(\varphi)\,d\varphi = \boldsymbol{0} \quad \text{and} \quad \int_0^{2\pi} \boldsymbol{h}(\varphi)\otimes\boldsymbol{h}(\varphi)\,d\varphi = \pi(\boldsymbol{I}-\boldsymbol{e}_1\otimes\boldsymbol{e}_1),$$

we find that

$$\boldsymbol{S}(z,0,0) = \Big(\iint d\boldsymbol{S}\Big)_{(z,0,0)} = s_1(z)\boldsymbol{e}_1\otimes\boldsymbol{e}_1 + s_2(z)(\boldsymbol{I}-\boldsymbol{e}_1\otimes\boldsymbol{e}_1),$$

where the proper values of tensor $\boldsymbol{S}(z,0,0)$, the one simple the other double, are:

$$s_1(z) := \frac{2\pi}{f}\int_0^{r_0} \widehat{\sigma}_1^B(z,s)\,ds = -p\left(1-\frac{z^3}{(r_0^2+z^2)^{\frac{3}{2}}}\right),$$

$$s_2(z) := \frac{\pi}{f}\int_0^{r_0} (\widehat{\sigma}_2^B(z,s)+\widehat{\sigma}_3^B(z,s))\,ds = -\frac{p}{2}\left(1+2\nu-\frac{2(1+\nu)z}{(r_0^2+z^2)^{\frac{1}{2}}}+\frac{z^3}{(r_0^2+z^2)^{\frac{3}{2}}}\right).$$

The *maximal tangential stress*—the stress parameter on which soil subsidence depends—is found for planes at an angle $\pi/4$ with the vertical axis; its value at z is:

$$\tau_{max}(z) = \frac{1}{2}|s_1(z)-s_2(z)|,$$

and its maximum is found at

$$z = \left(\frac{2(1+\nu)}{7-2\nu}\right)^{\frac{1}{2}} r_0 \approx \frac{2}{3}r_0.$$

Remark 5.15 It is not difficult to check that

$$\lim_{r_0\to 0+} s_1(z) = \widehat{\sigma}_1(z,0), \quad \lim_{r_0\to 0+} s_2(z) = \widehat{\sigma}_2(z,0) = \widehat{\sigma}_3(z,0), \quad \widehat{\sigma}_4(z,0) = 0,$$

and hence that

$$\lim_{r_0\to 0+} \boldsymbol{S}(z,0,0) = \boldsymbol{S}^B(z,0,0).$$

5.7.4 Horizontal Surface Displacements in the Flamant Problem

We wish to show how to evaluate the horizontal component at the surface of the Flamant displacement starting from the corresponding information for the Boussinesq Problem, that is, starting from

$$\breve{\boldsymbol{u}}_h^B(r,\varphi) = -\frac{f}{4\pi G}(1-2\nu)\,r^{-1}\widehat{\boldsymbol{h}}(\varphi) \tag{5.75}$$

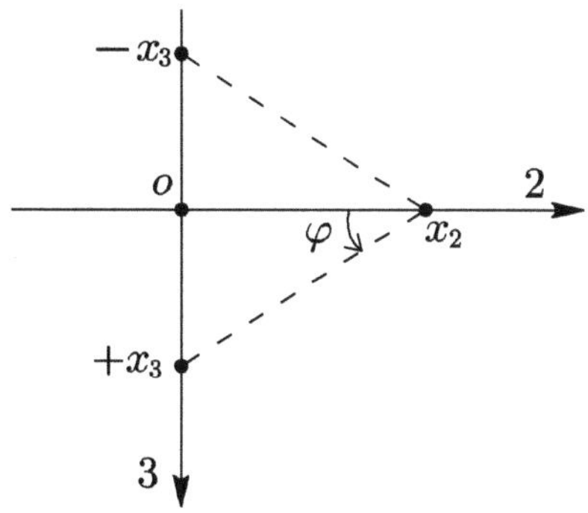

Fig. 5.11 A point of the x_2 axis where the effects of loads concentrated at points $(0, 0, \pm x_3)$ are superimposed

(cf. (5.72)). With reference to Fig. 5.11, we imagine to superimpose at point $(x_2, 0)$ the effects of loads $f\,dx_3$ concentrated at points of abscissa $\pm x_3$ of the Flamant load line. With the use of (5.75), we find that:

$$d\breve{\boldsymbol{u}}_h^F = -(f\,dx_3)\frac{1-2\nu}{4\pi G}r^{-1}\big(\widehat{\boldsymbol{h}}(\varphi) + \boldsymbol{h}(-\varphi)\big),$$

whence, given that

$$\widehat{\boldsymbol{h}}(\varphi) + \boldsymbol{h}(-\varphi) = 2\cos\varphi\,\boldsymbol{e}_2, \quad \cos\varphi = \frac{x_2}{\widehat{r}(x_3)}, \quad \widehat{r}^2(x_3) = x_2^2 + x_3^2,$$

we end up with

$$\breve{\boldsymbol{u}}_h^F(x_2) = -f\frac{1-2\nu}{2\pi G}x_2\Big(\int_0^{+\infty}\frac{1}{x_2^2+x_3^2}\,dx_3\Big)\boldsymbol{e}_2 = -f\frac{1-2\nu}{4G}\,\mathrm{sgn}\,(x_2)\boldsymbol{e}_2,$$

that is, to within a change of elastic moduli based on (2.55) and (2.56), the horizontal surface displacement expected on the basis of relation (4.35) for the Flamant Problem[13].

5.8 An Alternative Representation of the Boussinesq Elastic State

A good old-fashioned manner to represent the Boussinesq elastic state is found *via* a general representation in terms of harmonic functions for the displacement solution of the Navier equation (2.40) in the absence of distance forces. This latter

[13] This result follows from the fact that

$$\int\frac{1}{x_2^2+x_3^2}\,dx_3 = |x_2|^{-1}\arctan\frac{x_3}{|x_2|}.$$

representation, that is named after Boussinesq himself, Piotr Fiodorovitch Papkovitch and Heinz Neuber (see Appendix A.5), is:

$$2G\boldsymbol{u} = \boldsymbol{\psi} - \frac{1}{4(1-\nu)}\nabla(\boldsymbol{x}\cdot\boldsymbol{\psi} + \varphi), \quad \boldsymbol{x} := x - o, \tag{5.76}$$

where both the scalar potential φ and the vector potential $\boldsymbol{\psi}$ are harmonic. On recalling formulas (5.68) and (5.69), it is not difficult to verify that the potentials fitting Boussinesq displacement field $\boldsymbol{u}^B$ are:

$$\boldsymbol{\psi}^B = \psi^B \boldsymbol{e}_1 = \frac{1}{2\pi\rho}\boldsymbol{e}_1, \quad \varphi^B = \frac{1-2\nu}{2\pi}\log(x_1+\rho), \quad \rho^2 := x_1^2 + x_2^2 + x_3^2.$$

Once $\boldsymbol{u}^B$ is expressed in terms of ψ^B and φ^B, a sequential use of the compatibility equation (2.9) and the constitutive law (2.46) yields $\boldsymbol{E}^B$ first, then $\boldsymbol{S}^B$. We record here for later use the form of component S_{11}^B in terms of ψ^B and φ^B:

$$S_{11}^B = \frac{1}{1-\nu}\left(2(1-\nu)\psi^B_{,1} - \varphi^B_{,11} - x_1\psi^B_{,11}\right). \tag{5.77}$$

References

1. Boussinesq J (1878) Équilibre d'élasticité d'un sol isotrope sans pesanteur, supportant différents poids. CR Acad Sci 86:1260–1263
2. Boussinesq J (1885) Application des Potentiels à l'Étude de l'Équilibre et du Mouvement des Solides Élastiques. Gauthiers-Villars, Paris
3. Boussinesq J (1888) Équilibre d'élasticité d'un solide sans pesanteur, homogène et isotrope, dont les parties profondes sont maintenues fixes, pendant que sa surface éprouve des pressions ou des déplacements connus, s'annullant hors d'une région restreinte où ils sont arbitraires. CR Acad Sci 106(1043–1048):1119–1123
4. Bowles JE (1996) Foundation analysis and design. Mcgraw-Hill, New York
5. Davis RO, Selvadurai APS (1996) Elasticity and geomechanics. Cambridge University Press, Cambridge
6. Favata A (2012) On the Kelvin problem. J Elast 109:189–204
7. Flamant A (1892) Sur la répartition des pressions dans un solide rectangulaire chargé transversalement. CR Acad Sci 114:1465–1468
8. Gurtin ME (1972) The linear theory of elasticity. In: Flügge S (ed) Handbuch der Physik, vol VIa/2. Springer, Berlin
9. Malvern LE (1969) Introduction to the mechanics of a continuous medium. Prentice-Hall, Englewood Cliffs
10. Selvadurai APS (2007) The analytical method in geomechanics. Appl Mech Rev 60:87–105
11. Timoshenko S, Goodier JN (1951) Theory of elasticity. McGraw-Hill, New York

Chapter 6
The Kelvin Problem

Lord Kelvin (William Thompson, 1824–1907) solved the problem that was later named after him in 1848 [6].[1] The problem consists in finding the equilibrium state of a linearly elastic, isotropic material body occupying the whole space and being subject to a point load (Fig. 6.1).

6.1 Solution by Juxtaposition

The plane version of Kelvin's problem we study in the next section is a problem formulated on a plane orthogonal to a uniform line load (Fig. 6.2). As far as the applied loads are concerned, both the Kelvin Problem and its plane version can be regarded as the *juxtaposition of two anti-mirror symmetric problems*: two Boussinesq problems in the case of the 3-D Kelvin Problem, either two Boussinesq-Flamant or two plane Cerruti problems in the case of the 2-D Kelvin problem (Fig. 6.3; the Cerruti Problem is treated in Chap. 8).

6.1.1 Continuity Conditions at Sutures

Unfortunately, superposition of elastic states does not yield the desired Kelvin state, because it does not guarantee a 'seamless suture' over the common boundary. For this, two continuity conditions should be satisfied pointwise, the one for the *traction* field, the other for the *displacement* field[2]:

$$[\![\boldsymbol{Sn}]\!] = \boldsymbol{0}, \quad [\![\boldsymbol{u}]\!] = \boldsymbol{0}^2.$$

[1] An exposition of Kelvin's solution tailored after Love's [2] is found in the Appendix, Sect. A.6.

[2] Consistent with definition (1.19), here $[\![\Psi]\!] := \Psi^+ - \Psi^-$ denotes the *jump* of the field Ψ at a suture plane, in terms of the limits $\Psi^\pm$ of Ψ when the point of interest is attained from one or the other part of that plane.

P. Podio-Guidugli and A. Favata, *Elasticity for Geotechnicians*, Solid Mechanics and Its Applications 204, DOI: 10.1007/978-3-319-01258-2_6,

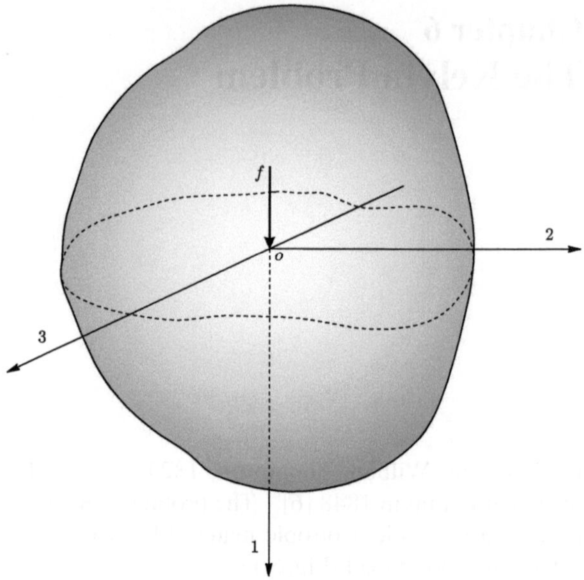

Fig. 6.1 The Kelvin Problem (this figure is taken from [1])

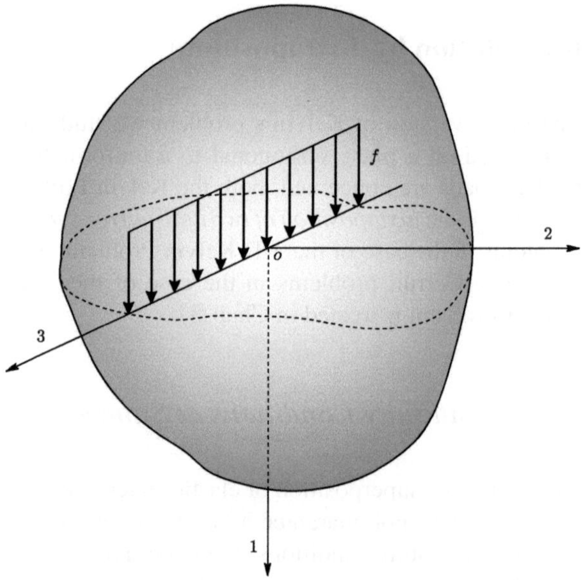

Fig. 6.2 The plane Kelvin problem

Juxtaposition of anti-mirror symmetric elastic states complies with the first condition trivially, because tractions are null all over the common boundary. On recalling the form of Flamant and Boussinesq displacement fields at $z = 0$, specified by, respectively, (4.35) and (5.72), we see that, while in both cases continuity of vertical displacements is gratis, horizontal components do jump:

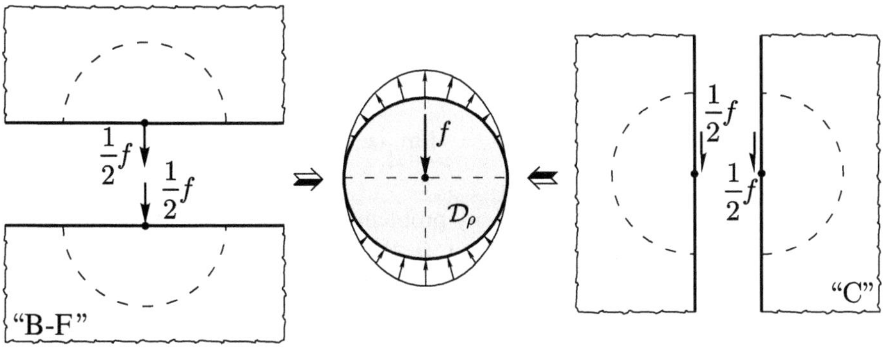

Fig. 6.3 Juxtaposition of loads and stress fields for two plane anti-mirror symmetric Boussinesq-Flamant and Cerruti Problems (this figure has been adapted from [4])

$$\left[\!\left[\boldsymbol{u}^F \cdot \boldsymbol{e}_2\right]\!\right]_{z=0} = -(1-2\nu)\frac{f}{E_0(1-\nu)}\operatorname{sgn} x_2\,,$$
$$\left[\!\left[\boldsymbol{u}^B \cdot \boldsymbol{h}\right]\!\right]_{z=0} = -(1-2\nu)\frac{f}{2\pi G}r^{-1};$$

thus, *solving the Kelvin Problem by juxtaposition is impossible.*

Nevertheless, we notice that, in both cases, continuity of horizontal components could be achieved for $\nu = 1/2$. This limit situation is excluded by the third of the inequalities (2.44), guaranteeing positivity of the elastic energy density stored by a *compressible* linearly elastic isotropic material. We see from (2.42) that, given the stress field and then $\operatorname{tr}\boldsymbol{S}$, the corresponding volume dilatation, measured by $\operatorname{tr}\boldsymbol{E}$, approaches zero when $\nu \to 1/2$, i.e., in the so-called *incompressibility limit*.[3] This fact prompts the expectation that, for incompressible linearly elastic materials, the Kelvin Problem be solvable by juxtaposition of two anti-mirror symmetric Boussinesq Problems for materials in the same class. We leave for the reader a task that is easy, after we solve the Boussinesq Problem for incompressible materials in the next subsection.

6.1.2 *Conditional Solvability: The Boussinesq Problem for Incompressible Materials*

An elasticity problem is solved when the relative *elastic state*—that is, the triplet $(\boldsymbol{u}, \boldsymbol{E}, \boldsymbol{S})$ of displacement, deformation and stress fields—is known. When an *internal constraint* prevails—that is, an *a priori* limitation on admissible deformations is posed—it would be desirable to deduce the elastic state from the elastic state of the

[3] We also see from (2.47) that, under the same circumstances, for the stored energy to stay finite the volume changes must become smaller and smaller as ν approaches $1/2$.

corresponding unconstrained problem. In the present case, incompressibility is the internal constraint we deal with, and we would like to give a precise meaning to the following formal writing:

$$(\boldsymbol{u}, \boldsymbol{E}, \boldsymbol{S})^{inc} = \lim_{\nu \to 1/2} (\boldsymbol{u}, \boldsymbol{E}, \boldsymbol{S}).$$

Now, the solution of a linear elasticity problem *depends with continuity on data*, that is, on the information we have about: (i) the geometry of the region on which the problem is formulated; (ii) the nature of the material filling that region; (iii) the applied loads; (iv) the boundary conditions. The value of the Poisson modulus is a datum, on which the solution depends in general with continuity, as it is possible to see, for instance, in (4.37) and (5.70). Therefore, it makes sense to expect that the displacement field for the incompressible Boussinesq Problem be obtained by taking the limit for $\nu \to 1/2$ of the same field in the compressible case, which is:

$$\boldsymbol{u}^B_{inc} = \frac{f}{4\pi G}\rho^{-1}((\cos^2\vartheta + 1)\boldsymbol{e}_1 + |\sin\vartheta|\cos\vartheta\,\boldsymbol{h}). \tag{6.1}$$

Moreover, given that the operations of taking a spatial gradient and the incompressibility limit commute, we have from (5.71) that

$$\begin{aligned}
(E^B_{inc})_{zz} &= \frac{f}{4\pi G}\rho^{-2}\cos\vartheta(-3\cos^2\vartheta + 1),\\
(E^B_{inc})_{rr} &= \frac{f}{4\pi G}\rho^{-2}\cos\vartheta(-3\sin^2\vartheta + 1),\\
(E^B_{inc})_{\varphi\varphi} &= \frac{f}{4\pi G}\rho^{-2}\cos\vartheta,\\
(E^B_{inc})_{zr} &= -\frac{3f}{4\pi G}\rho^{-2}\cos^2\vartheta|\sin\vartheta|.
\end{aligned}$$

It is easily checked that

$$\operatorname{tr}\boldsymbol{E}^B_{inc} = 0;$$

thus, the strain field $\boldsymbol{E}^B_{inc}$ is deviatoric:[4]

$$\boldsymbol{E}^B_{inc} = \operatorname{dev}\boldsymbol{E}^B_{inc}.$$

On recalling that $\cos\vartheta = \rho^{-1}z$, it is equally easy to see that, for $z = 0$,

[4] Needless to say, the same developments follow by an application of definition $(2.2)_2$ to the field (6.1). Recall that each symmetric tensor $\boldsymbol{A}$ can be additively split into uniquely defined *deviatoric* and *spheric* parts:

$$\boldsymbol{A} = \operatorname{dev}\boldsymbol{A} + \operatorname{sph}\boldsymbol{A}, \quad \operatorname{sph}\boldsymbol{A} := \frac{1}{3}\operatorname{tr}\boldsymbol{A}, \quad \operatorname{dev}\boldsymbol{A} := \boldsymbol{A} - \operatorname{sph}\boldsymbol{A}.$$

$$\boldsymbol{E}_{inc}^{B}(0,r)\equiv\boldsymbol{0},\tag{6.2}$$

Finding $\boldsymbol{S}^{inc}$ requires something more than taking a limit: an *ad hoc* modeling assumption is needed.

The constitutive Eq. $(2.46)_2$ for a compressible isotropic material can be written as follows:

$$\boldsymbol{S}=2G\Big(\operatorname{dev}\boldsymbol{E}+\frac{1+\nu}{1-2\nu}\operatorname{sph}\boldsymbol{E}\Big).\tag{6.3}$$

In the incompressibility limit, the value of G is kept fixed, while both $(1-2\nu)$ and $\operatorname{sph}\boldsymbol{E}$ tend to null; it is then necessary to give the limit of $(1-2\nu)^{-1}\operatorname{sph}\boldsymbol{E}$ a meaning. We assume that a finite limit exists:

$$\lim_{\nu\to 1/2}\frac{1+\nu}{1-2\nu}\operatorname{sph}\boldsymbol{E}=\pi\boldsymbol{I},$$

with the scalar-valued field π *constitutively indetermined*. Accordingly, we replace (6.3) by the constitutive equation:

$$\boldsymbol{S}=2G\operatorname{dev}\boldsymbol{E}+\pi\boldsymbol{I},$$

describing the mechanical response of a incompressible isotropic material, and we write, provisionally,

$$\boldsymbol{S}_{inc}^{B}=2G\operatorname{dev}\boldsymbol{E}_{inc}^{B}+\pi\boldsymbol{I}.$$

The equilibrium pressure field is determined by requiring that the stress field $\boldsymbol{S}_{inc}^{B}$ be divergenceless in the interior of $\mathcal{HS}^{+}$, a condition that reads:

$$\nabla\widehat{\pi}(z,r)=-2G\operatorname{div}\widehat{\boldsymbol{E}}_{inc}^{B}(z,r)\quad\text{for } z,r>0,$$

and by satisfying the boundary condition (5.15), which, in view of (6.2), reduces to:

$$\pi(0,r)=0,\quad r>0.$$

Remark 6.1 A material is *constrained* whenever some deformations are deemed constitutively impossible by requesting that the strain measure $\boldsymbol{E}$ satisfy an algebraic limitation of the following type:

$$\boldsymbol{V}\cdot\boldsymbol{E}=0,\tag{6.4}$$

for a given *constraint tensor* $\boldsymbol{V}\in\mathrm{Sym}$. For a constrained material, it is customary to decompose the stress tensor additively:

$$\boldsymbol{S}=\boldsymbol{S}^{(A)}+\boldsymbol{S}^{(R)},$$

with the *active stress* $\mathbf{S}^{(A)}$ determined by a tensor-valued constitutive function, defined on $\mathcal{A} := \{\boldsymbol{E} \mid \boldsymbol{V}\cdot\boldsymbol{E} = 0\}$, the subspace of Sym composed by all admissible deformations, and with the *reactive stress* $\mathbf{S}^{(R)}$ (i.e., the stress necessary to maintain the stipulated kinematic constraint), characterized by the condition that the work spent on whatever admissible deformation be null:

$$\mathbf{S}^{(R)}\cdot\boldsymbol{E} = 0, \quad \forall\boldsymbol{E}\in\mathcal{A}.$$

This last condition is equivalent to the following representation of the reactive stress:

$$\mathbf{S}^{(R)} = \sigma^{(R)}\boldsymbol{V},$$

where $\sigma^{(R)}$ a constitutively indeterminate scalar multiplier (e.g., for $\boldsymbol{V} = \boldsymbol{I}$, $\sigma^{(R)} = \pi$).[5]

Remark 6.2 The response symmetry of a constrained material is affected by the nature of the internal constraints, if any. The internal constraints compatible with isotropy are three: two are nontrivial, incompressibility and *shape preservation*, for which it is required that $\operatorname{dev}\boldsymbol{E} = \boldsymbol{0}$; one is trivial, *rigidity*, in which case the choice of a constraint tensor in (6.4) is arbitrary, and hence $\boldsymbol{E} = \boldsymbol{0}$; for a rigid material, the active stress is null, all stress is of reactive nature.

6.2 The 2-D Kelvin Problem

Suppose that a constant line load $\boldsymbol{f} = f\boldsymbol{e}_1$ (with $\dim(f) = FL^{-1}$) is applied along the x_3-axis (Fig. 6.2). To find the relative equilibrium state, our plan is:
(i) to individuate a large class of two-dimensional balanced stress fields, that is to say, stress fields of the form (4.8) that solve the *distributional equilibrium equation*

$$\operatorname{div}\mathbf{S}(x) + f\delta(o)\boldsymbol{e}_1 = \boldsymbol{0} \quad \text{for } x\in\mathcal{H}; \tag{6.5}$$

(ii) to add to each of such stress fields an auxiliary stress field:

$$\mathbf{S}^{(aux)} = S_{33}\boldsymbol{e}_3\otimes\boldsymbol{e}_3, \quad S_{33} = \nu(S_{11}+S_{22}), \tag{6.6}$$

so as to obtain a family of three-dimensional stress fields, among which to choose, by means of condition (2.69), those compatible with the existence of a state of plane strain and deformation in the whole space;
(iii) to construct such strain and deformation states.

[5] More about internal constraint in linear elasticity is found in [3], Chapter III, Sections 17 and 18.

6.2.1 Balanced Stress Fields

We recall that a locally integrable stress field $\boldsymbol{S}$ being divergenceless over $\mathcal{H} \setminus o$ is said to solve equation (6.5) *in the sense of distributions over* $\mathcal{H}$ if

$$\int_{\mathcal{H}} \boldsymbol{S} \cdot \nabla \boldsymbol{v} = f \boldsymbol{v}(o) \cdot \boldsymbol{e}_1 \quad \text{for all test vector fields } \boldsymbol{v} \in C_c^\infty(\mathcal{H}, \mathcal{V}^{(2)}); \qquad (6.7)$$

here, as the notation suggests, a test vector field is a C^∞ field with compact support, defined over $\mathcal{H}$ and taking its values in the 2-D vector space $\mathcal{V}^{(2)}$, the translation space of $\mathcal{H}$. The direct mechanical interpretation of a condition of this type is that, for a stress field to balance the applied loads, *the stress working must equal the load working, for whatever test velocity field.*[6] We shall now derive a version of this condition that allows for a different and more specific mechanical interpretation.

For each fixed test field $\boldsymbol{v}$, let $\mathcal{D}_\rho$ be a disk of radius ρ centered at o and containing the support of $\boldsymbol{v}$, and let $\mathcal{D}_\varepsilon$ be a smaller disk, also centered at o. Then,

$$\int_{\mathcal{H}} \boldsymbol{S} \cdot \nabla \boldsymbol{v} = \int_{\mathcal{D}_\rho} \boldsymbol{S} \cdot \nabla \boldsymbol{v} = \int_{\mathcal{D}_\varepsilon} \boldsymbol{S} \cdot \nabla \boldsymbol{v} + \int_{\mathcal{D}_\rho \setminus \mathcal{D}_\varepsilon} \boldsymbol{S} \cdot \nabla \boldsymbol{v}.$$

Given that $\boldsymbol{S}$ is integrable and $\nabla \boldsymbol{v}$ is smooth,

$$\lim_{\varepsilon \to 0} \int_{\mathcal{D}_\varepsilon} \boldsymbol{S} \cdot \nabla \boldsymbol{v} = 0.$$

Moreover, in view of the identity

$$\boldsymbol{S} \cdot \nabla \boldsymbol{v} = \operatorname{div}(\boldsymbol{S}^T \boldsymbol{v}) - \boldsymbol{v} \cdot \operatorname{div} \boldsymbol{S},$$

the divergence theorem, the fact that $\operatorname{supp}(\boldsymbol{v}) \subset \mathcal{D}_\rho$, and the fact that $\operatorname{div} \boldsymbol{S}$ is null over $\mathcal{H} \setminus o$, we have that

$$\int_{\mathcal{D}_\rho \setminus \mathcal{D}_\varepsilon} \boldsymbol{S} \cdot \nabla \boldsymbol{v} = \int_{\partial(\mathcal{D}_\rho \setminus \mathcal{D}_\varepsilon)} \boldsymbol{S}\boldsymbol{n} \cdot \boldsymbol{v} - \int_{\mathcal{D}_\rho \setminus \mathcal{D}_\varepsilon} \boldsymbol{v} \cdot \operatorname{div} \boldsymbol{S} = - \int_{\partial \mathcal{D}_\varepsilon} \boldsymbol{S}\boldsymbol{n} \cdot \boldsymbol{v}.$$

Therefore, for each admissible test field, condition (6.7) can be given the provisional form

$$\lim_{\varepsilon \to 0} \left(\int_{\partial \mathcal{D}_\varepsilon} \boldsymbol{S}\boldsymbol{n} \cdot \boldsymbol{v} \right) + f \boldsymbol{v}(o) \cdot \boldsymbol{e}_1 = 0.$$

Note that

[6] Alternative terminological choices are 'power' (or 'power expenditure') for 'working' and 'virtual' for 'test'; an alternative version of the italicized sentence above would read: *the stress power equals the load power for whatever virtual velocity field.*

$$\int_{\partial \mathcal{D}_\varepsilon} \boldsymbol{Sn} \cdot \boldsymbol{v} = \int_{-\pi}^{+\pi} \varepsilon \boldsymbol{S}(o + \varepsilon \widehat{\boldsymbol{e}}(\vartheta)) \widehat{\boldsymbol{e}}(\vartheta) \cdot \boldsymbol{v}(o + \varepsilon \widehat{\boldsymbol{e}}(\vartheta))\, d\vartheta.$$

Thus, if the following condition holds:
(A) the vector field $\vartheta \mapsto \varepsilon \widehat{\boldsymbol{S}}(o + \varepsilon \widehat{\boldsymbol{e}}(\vartheta)) \widehat{\boldsymbol{e}}(\vartheta)$ is independent of ε,
then

$$\lim_{\varepsilon \to 0} \Big(\int_{\partial \mathcal{D}_\varepsilon} \boldsymbol{Sn} \cdot \boldsymbol{v} \Big) = \Big(\int_{-\pi}^{+\pi} \varepsilon \widehat{\boldsymbol{S}}(\varepsilon, \vartheta) \widehat{\boldsymbol{e}}(\vartheta)\, d\vartheta \Big) \cdot \boldsymbol{v}(o),$$

and (6.7) can be given the final form

$$\int_{\partial \mathcal{D}_\varepsilon} \boldsymbol{Sn} + f \boldsymbol{e}_1 = \boldsymbol{0}.$$

The mechanical interpretation of this condition on the stress field—that the diffused contact force over the periphery of any disk balances the concentrated force applied at its center—can be seen as a counterpart of the mathematical interpretation of condition (6.5)—that the corresponding balanced stress field has *divergence measure* supported at the point where the concentrated force is applied.

It is not difficult to see that each stress field of the one-parameter family

$$\begin{aligned} \boldsymbol{S} = \widehat{\boldsymbol{S}}(\rho, \vartheta; \boldsymbol{e}_1) = \rho^{-1} (&\alpha_0 \cos\vartheta\, \widehat{\boldsymbol{e}}(\vartheta) \otimes \widehat{\boldsymbol{e}}(\vartheta) + \gamma_0 \cos\vartheta\, \widehat{\boldsymbol{e}}'(\vartheta) \otimes \widehat{\boldsymbol{e}}'(\vartheta) \\ &+ \gamma_0 \sin\vartheta (\widehat{\boldsymbol{e}}(\vartheta) \otimes \widehat{\boldsymbol{e}}'(\vartheta) + \widehat{\boldsymbol{e}}'(\vartheta) \otimes \widehat{\boldsymbol{e}}(\vartheta))), \quad \alpha_0 - \gamma_0 = -\frac{f}{\pi}, \end{aligned} \tag{6.8}$$

fulfills condition (A) and balances the applied load. In particular, the second of (6.8) follows from the balance of a body part in the form of a disk centered at the origin, of arbitrary radius ρ: since

$$\rho \boldsymbol{Se} = \alpha_0 \cos\vartheta\, \boldsymbol{e} + \gamma_0 \sin\vartheta\, \boldsymbol{e}', \tag{6.9}$$

an easy calculation shows that

$$\int_{\partial \mathcal{D}_\rho} \boldsymbol{Se} = -f \boldsymbol{e}_1 \tag{6.10}$$

(cf. e.g., [5], Section 78).

Remark 6.3 With the use of $(6.8)_2$, it is not difficult to transform $(6.9)_2$ into

$$\widehat{\boldsymbol{S}}(\rho, \vartheta; \boldsymbol{e}_1) \boldsymbol{e}(\vartheta) = \rho^{-1} \Big(\alpha_0 \widehat{\boldsymbol{e}}(2\vartheta) + \frac{f}{\pi} \sin\vartheta\, \widehat{\boldsymbol{e}}'(\vartheta) \Big),$$

which allows for an easier visualization of the stress vector at any point of $\partial \mathcal{D}_\rho$; note that the first addendum does not contribute to the integral in (6.10).

Remark 6.4 The stress fields (6.8) have the form (4.8). Condition (4.9) has been dropped, because it makes no sense for the full-plane domain where Kelvin problem is formulated. The choices of $\widehat{a}$ and $\widehat{c}$ reflect the expected parities of these two functions. Choosing $\widehat{a}(\vartheta) = \sin\vartheta = \widehat{c}(\vartheta)$ leads to the Kelvin stress fields for the load $\boldsymbol{f} = f\boldsymbol{e}_2$, namely,

$$\widehat{\boldsymbol{S}}(\rho, \vartheta; \boldsymbol{e}_2) = \rho^{-1}(\alpha_0 \sin\vartheta\, \widehat{\boldsymbol{e}}(\vartheta) \otimes \widehat{\boldsymbol{e}}(\vartheta) - \gamma_0 \sin\vartheta\, \widehat{\boldsymbol{e}}'(\vartheta) \otimes \widehat{\boldsymbol{e}}'(\vartheta)$$
$$+ \gamma_0 \cos\vartheta(\widehat{\boldsymbol{e}}(\vartheta) \otimes \widehat{\boldsymbol{e}}'(\vartheta) + \widehat{\boldsymbol{e}}'(\vartheta) \otimes \widehat{\boldsymbol{e}}(\vartheta))), \quad \alpha_0 + \gamma_0 = -\frac{f}{\pi}.$$

6.2.2 Compatible Stress Fields

As anticipated, we now seek what stress fields of the type (6.8) satisfy the compatibility condition (2.69). This is quickly done. Firstly, from (6.8) we deduce that

$$\operatorname{tr}\boldsymbol{S} = (\alpha_0 + \gamma_0)\rho^{-1}\cos\vartheta.$$

Then, with the use of the last of (3.19), we find that

$$\Delta(\rho^{-1}\cos\vartheta) = 0.$$

We then conclude, by taking (6.6) into account, that each of the stress fields of the one-parameter family

$$\widetilde{\boldsymbol{S}} = \boldsymbol{S} + \nu(\operatorname{tr}\boldsymbol{S})\boldsymbol{e}_3 \otimes \boldsymbol{e}_3,$$

is compatible with a state of plane strain and plane displacement, to be determined in the next subsection.

6.2.3 Strain and Displacements Fields

The strain field solving the plane Kelvin problem is obtained by inserting the stress field (6.8) into the inverse constitutive equation (2.57). One finds:

$$\boldsymbol{E} = \frac{1}{E_0}\rho^{-1}((\alpha_0 - \nu_0\gamma_0)\cos\vartheta\, \boldsymbol{e} \otimes \boldsymbol{e} + (\gamma_0 - \nu_0\alpha_0)\cos\vartheta\, \boldsymbol{e}' \otimes \boldsymbol{e}'$$
$$+ (1 + \nu_0)\gamma_0 \sin\vartheta(\boldsymbol{e} \otimes \boldsymbol{e}' + \boldsymbol{e}' \otimes \boldsymbol{e})).$$

To determine the displacement field, one has to solve the following system of PDEs:

$$\begin{aligned} u_{\rho,\rho} &= \rho^{-1}\frac{\alpha_0 - \nu_0\gamma_0}{E_0}\cos\vartheta, \\ u_{\vartheta,\vartheta} + u_\rho &= \frac{\gamma_0 - \nu_0\alpha_0}{E_0}\cos\vartheta, \\ u_{\vartheta,\rho} + \rho^{-1}(u_{\rho,\vartheta} - u_\vartheta) &= \rho^{-1}\frac{2(1+\nu_0)}{E_0}\gamma_0\sin\vartheta, \end{aligned} \tag{6.11}$$

where the unknown fields

$$u_\rho := \boldsymbol{u}\cdot\boldsymbol{e} = \hat{u}_\rho(\rho,\vartheta), \qquad u_\vartheta := \boldsymbol{u}\cdot\boldsymbol{e}' = \hat{u}_\vartheta(\rho,\vartheta),$$

must satisfy the intrinsic symmetry conditions of the plane Kelvin problem and therefore be such that

$$\hat{u}_\rho(\rho,\vartheta) = \hat{u}_\rho(\rho,-\vartheta), \qquad \hat{u}_\vartheta(\rho,\vartheta) = -\hat{u}_\vartheta(\rho,-\vartheta). \tag{6.12}$$

The integration of $(6.11)_1$ yields:

$$\hat{u}_\rho(\rho,\vartheta) = \frac{\alpha_0 - \nu_0\gamma_0}{E_0}\log\rho\cos\vartheta + \hat{v}(\vartheta), \tag{6.13}$$

with $\hat{v}$ an arbitrary even function, so as to satisfy condition $(6.12)_1$. With this provisional representation for $\hat{u}_\rho$, integration of $(6.11)_2$ yields:

$$\hat{u}_\vartheta(\rho,\vartheta) = -\frac{\alpha_0 - \nu_0\gamma_0}{E_0}\log\rho\sin\vartheta - \widehat{V}(\vartheta) + \frac{\gamma_0 - \nu_0\alpha_0}{E_0}\sin\vartheta, \tag{6.14}$$

where $\widehat{V}$ is a primitive of $\hat{v}$, and hence is odd. The addition of an arbitrary function of ρ to this expression of $\hat{u}_\vartheta$ is forbidden by condition $(6.12)_2$. Moreover, the third of (6.11) determines $\hat{v}$: on inserting (6.13) and (6.14) into it, we find that

$$-\frac{\alpha_0 - \nu_0\gamma_0}{E_0}\sin\vartheta + \hat{v}'(\vartheta) + \widehat{V}(\vartheta) - \frac{\gamma_0 - \nu_0\alpha_0}{E_0}\sin\vartheta = \frac{2(1+\nu_0)}{E_0}\gamma_0\sin\vartheta,$$

or rather, after differentiation and term rearrangement,

$$\hat{v}''(\vartheta) + \hat{v}(\vartheta) = \frac{(3+\nu_0)\gamma_0 + (1-\nu_0)\alpha_0}{E_0}\cos\vartheta.$$

The even solutions of this equation are:

$$\hat{v}(\vartheta) = v_0\cos\vartheta + \frac{1}{2E_0}((3+\nu_0)\gamma_0 + (1-\nu_0)\alpha_0)\vartheta\sin\vartheta;$$

their primitives are:

$$\widehat{V}(\vartheta) = v_0 \sin\vartheta - \frac{1}{2E_0}((3+\nu_0)\gamma_0 + (1-\nu_0)\alpha_0)(\vartheta\cos\vartheta - \sin\vartheta). \tag{6.15}$$

In addition to the parity requirements specified by (6.12), the displacement field must obey the 'glueing condition':

$$\boldsymbol{u}(\rho, -\pi) = \boldsymbol{u}(\rho, +\pi),$$

which, upon fiddling a bit with relations (6.13)–(6.15), is found equivalent to the scalar condition $\hat{V}(\pi) = 0$, or rather:

$$(3+\nu_0)\gamma_0 + (1-\nu_0)\alpha_0 = 0;$$

together with $(6.8)_2$, this condition allows to determine the two constants α_0 and γ_0:

$$\alpha_0 = -\frac{f}{4\pi}(3+\nu_0), \quad \gamma_0 = \frac{f}{4\pi}(1-\nu_0).$$

In conclusion, the plane Kelvin problem is solved by the displacement field:

$$\boldsymbol{u} = \hat{u}_\rho(\rho, \vartheta)\boldsymbol{e}(\vartheta) + \hat{u}_\vartheta(\rho, \vartheta)\boldsymbol{e}'(\vartheta),$$

with

$$\begin{aligned}
\hat{u}_\rho(\rho, \vartheta) &= \frac{f}{4\pi\, E_0}\,(3+\nu_0^2)\,\log\rho\cos\vartheta, \\
\hat{u}_\vartheta(\rho, \vartheta) &= \frac{f}{4\pi\, E_0}\big(-(3+\nu_0^2)\log\rho + 1 + \nu_0 + 3\nu_0^2\big)\sin\vartheta
\end{aligned}$$

(we have disposed of the rigid displacement:

$$\boldsymbol{u}_{rig} = v_0\left(\cos\vartheta\boldsymbol{e}(\vartheta) - \sin\vartheta\boldsymbol{e}'(\vartheta)\right) = v_0\,\boldsymbol{e}_1$$

by setting to null the constant v_0); the corresponding stress field is:

$$\begin{aligned}
\boldsymbol{S} = {} & S_{\rho\rho}(\rho, \vartheta)\widehat{\boldsymbol{e}}(\vartheta)\otimes\widehat{\boldsymbol{e}}(\vartheta) + S_{\vartheta\vartheta}(\rho, \vartheta)\widehat{\boldsymbol{e}}'(\vartheta)\otimes\widehat{\boldsymbol{e}}'(\vartheta) \\
& + S_{\rho\vartheta}(\rho, \vartheta)\big(\widehat{\boldsymbol{e}}(\vartheta)\otimes\widehat{\boldsymbol{e}}'(\vartheta) + \widehat{\boldsymbol{e}}'(\vartheta)\otimes\widehat{\boldsymbol{e}}(\vartheta)\big),
\end{aligned} \tag{6.16}$$

with

$$\begin{aligned}
S_{\rho\rho} &= -\frac{f}{4\pi}(3+\nu_0)\rho^{-1}\cos\vartheta, \\
S_{\vartheta\vartheta} &= \frac{f}{4\pi}(1-\nu_0)\rho^{-1}\cos\vartheta, \\
S_{\rho\vartheta} &= \frac{f}{4\pi}(1-\nu_0)\rho^{-1}\sin\vartheta.
\end{aligned} \tag{6.17}$$

Remark 6.5 In the expression (4.37) for the displacement field of the Flamant Problem, there is a term proportional to $\vartheta \cos \vartheta$, that is, of the same kind of the term we just eliminated by imposing the 'glueing condition'. Actually, in that problem, this condition does not apply, because all (displacement, strain, stress) fields are defined for ϑ variable in the interval $[-\pi/2, +\pi/2]$. This remark prompts us to underline a relevant difference in the posing of the Boussinesq-Flamant Problem and the plane Kelvin Problem. Although the equilibrium equations are the same, the domains on which the two problems are formulated are different. On the one hand, the need to satisfy the boundary conditions prevailing on the plane $z = 0$ reduces the class of balanced and compatible stress fields for the Flamant Problem to a subclass of that for the Kelvin Problem; on the other hand, in the latter problem, the larger freedom in the choice of stress fields is compensated by an additional kinematic constraint, the glueing condition, allowing for the determination of the unique solution.

6.3 The Kelvin Elastic State

6.3.1 The Stress Field

The Kelvin Problem is similar to Boussinesq's in that it enjoys the same cylindrical symmetry. Once again system (5.31) must be solved for a compatible stress field of the form (5.12):

$$S = \sigma_1 \boldsymbol{e}_1 \otimes \boldsymbol{e}_1 + \sigma_2 \boldsymbol{h} \otimes \boldsymbol{h} + \sigma_3 \boldsymbol{h}' \otimes \boldsymbol{h}' + \sigma_4 (\boldsymbol{e}_1 \otimes \boldsymbol{h} + \boldsymbol{h} \otimes \boldsymbol{e}_1),$$

with the sequential procedure introduced on Sect. 5.3.3; in particular, the first three steps of that procedure allow to determine the expressions for the stress trace and the stress components σ_1, and σ_4, that we here recall for the reader's convenience:

$$\begin{aligned} &\widetilde{\alpha}(\rho, \vartheta) = \alpha_0 \, \rho^{-2} \cos \vartheta, \\ &\widetilde{\sigma}_1(\rho, \vartheta) = \rho^{-2} \left(\widetilde{\tau}_1(\vartheta) + \beta_0 \cos \vartheta \right), \quad \widetilde{\tau}_1(\vartheta) = \frac{3}{2} \alpha_0 \cos^3 \vartheta \\ &\widetilde{\sigma}_4(\rho, \vartheta) = \rho^{-2} \left(\widetilde{\tau}_4(\vartheta) + \beta_0 |\sin \vartheta| \right), \quad \widetilde{\tau}_4(\vartheta) = \frac{3}{2} \alpha_0 \cos^2 \vartheta \, |\sin \vartheta|. \end{aligned} \tag{6.18}$$

What makes the difference are the values to assign to constants α_0, β_0. We begin to gain information on this point by imposing that a ball centered at the origin be in equilibrium:

$$f = -2\pi \int_0^{\pi} (\cos \vartheta \, \widetilde{\tau}_1(\vartheta) + |\sin \vartheta| \, \widetilde{\tau}_4(\vartheta)) |\sin \vartheta| \, d\vartheta,$$

whence

$$\alpha_0 + 2\beta_0 = -\frac{f}{2\pi}\,. \tag{6.19}$$

We also record an alternative way of writing (6.18):

$$\begin{aligned}
\widehat{\alpha}(z,r) &= \alpha_0 \frac{z}{\rho^3},\\
\widehat{\sigma}_1(z,r) &= \frac{3}{2}\alpha_0 \frac{z^3}{\rho^5} + \beta_0 \frac{z}{\rho^3},\\
\widehat{\sigma}_4(z,r) &= \frac{3}{2}\alpha_0 \frac{z^2 r}{\rho^5} + \beta_0 \frac{r}{\rho^3}.
\end{aligned} \tag{6.20}$$

The stress components σ_2 and σ_3 can be determined in the same way as for the Boussinesq Problem. To take step 4 (Sect. 5.4.3), we replace (5.56) by

$$\sigma_2 = -\sigma_3 - \sigma_1 + (1+\nu)\alpha = -\sigma_3 - \frac{3}{2}\alpha_0 \frac{z^3}{\rho^5} + (\alpha_0(1+\nu) - \beta_0)\frac{z}{\rho^3}, \tag{6.21}$$

with which the differential Eq. (5.57) is replaced by:

$$\sigma_3 + (r\sigma_3)_{,r} = -\frac{3}{2}\alpha_0 \frac{z^3}{\rho^5} + (\alpha_0(1+\nu) - \beta_0)\frac{z}{\rho^3} + 3\alpha_0\,\nu\,\frac{r^2 z}{\rho^5},$$

whose solution is:

$$\widehat{\sigma}_3(z,r) = -\big(\alpha_0(1-2\nu) - \beta_0\big)\frac{z}{\rho^3} - \big(\alpha_0(1-2\nu) - 2\beta_0\big)\frac{z^3}{2r^2\rho^3} + \frac{g(z)}{r^2}\,;$$

combining this with $(6.21)_2$, we also have that

$$\begin{aligned}
\widehat{\sigma}_2(z,r) = &-\frac{3}{2}\alpha_0 \frac{z^3}{\rho^5} + (\alpha_0(1+\nu) - \beta_0)\frac{z}{\rho^3}\\
&+ \big(\alpha_0(1-2\nu) - \beta_0\big)\frac{z}{\rho^3} + \big(\alpha_0(1-2\nu) - 2\beta_0\big)\frac{z^3}{2r^2\rho^3} - \frac{g(z)}{r^2}
\end{aligned}$$

(cf. the last two equations in Sect. 3.4 of [1]). It remains for us to complete the determination of constants α_0, β_0, and to find the form of function g. We do it in a manner completely similar to what we did for the same purpose in Sect. 5.6.

Firstly, by using the inverse constitutive law $(2.45)_2$ and (6.20), we find that:

$$E_{\varphi\varphi} = -\frac{1}{2G}\left(\big(\alpha_0(1-\nu) - \beta_0\big)\frac{z}{\rho^3} + \big(\alpha_0(1-2\nu) - 2\beta_0\big)\frac{z^3}{2r^2\rho^3} - \frac{g(z)}{r^2}\right).$$

Secondly, with this and $(5.66)_3$, we obtain the following provisional expression for the radial displacement of points on any chosen horizontal plane:

$$u_r = rE_{\varphi\varphi} = -\frac{1}{2G}\left(\left(\alpha_0(1-\nu)-\beta_0\right)\frac{zr}{\rho^3} + \left(\alpha_0(1-2\nu)-2\beta_0\right)\frac{z^3}{2r\rho^3} - \frac{g(z)}{r}\right). \tag{6.22}$$

Thirdly, we impose again the kinematic symmetry condition (5.60):

$$\lim_{r\to 0^+} u_r(z,r) = 0,$$

and deduce from it that:

$$\alpha_0(1-2\nu)-2\beta_0 = 0, \quad g(z) = 0; \tag{6.23}$$

relations (6.19) and $(6.23)_1$ imply that:

$$\alpha_0 = -\frac{f}{4\pi(1-\nu)}, \qquad \beta_0 = -\frac{f(1-2\nu)}{8\pi(1-\nu)}. \tag{6.24}$$

In conclusion, the Kelvin stress components turn out to have the following expressions:

$$\begin{aligned}
\widehat{\sigma}_1^K(z,r) &= -\frac{f}{8\pi(1-\nu)}\left(3\frac{z^3}{\rho^5} - (1-2\nu)\frac{z}{\rho^3}\right),\\
\widehat{\sigma}_2^K(z,r) &= -\frac{f}{8\pi(1-\nu)}\left(3\frac{zr^2}{\rho^5} + (1-2\nu)\frac{z}{\rho^3}\right),\\
\widehat{\sigma}_3^K(z,r) &= \frac{f(1-2\nu)}{8\pi(1-\nu)}\frac{z}{\rho^3},\\
\widehat{\sigma}_4^K(z,r) &= -\frac{f}{8\pi(1-\nu)}\left(3\frac{z^2r}{\rho^5} - (1-2\nu)\frac{z}{\rho^3}\right)
\end{aligned} \tag{6.25}$$

(cf. Equations (40) in [1]).

6.3.2 The Strain and Displacement Fields

To deduce the strain field in Kelvin's problem, we combine the inverse constitutive Eq. (2.45) with (6.25), and find:

$$
\begin{aligned}
\widehat{E}^K_{zz}(z,r) &= -\frac{f}{16\pi G(1-\nu)\rho^5}\left(4(1+\nu)z^3 + (1-4\nu)zr^2\right),\\
\widehat{E}^K_{rr}(z,r) &= \frac{f}{16\pi G(1-\nu)\rho^5}\left(z^3 - 2zr^2\right),\\
\widehat{E}^K_{\varphi\varphi}(z,r) &= \frac{f}{16\pi G(1-\nu)}\frac{z}{\rho^3},\\
\widehat{E}^K_{zr}(z,r) &= -\frac{f}{16\pi G(1-\nu)\rho^5}\left(2(2-\nu)z^2 r + (1-2\nu)r^3\right)
\end{aligned}
\tag{6.26}
$$

(cf. equations (41) in [1]). As to the displacement field, it is the matter of a straightforward calculation to substitute (6.24) into (6.22), to obtain, in view also of $(6.23)_2$, that

$$
\hat{u}^K_r(z,r) = \frac{f}{16\pi G(1-\nu)}\frac{zr}{\rho^3}. \tag{6.27}
$$

Moreover, $(5.66)_1$ and $(6.26)_1$ imply that

$$
u_z = \frac{f}{16\pi G(1-\nu)}\left(\frac{2(1-2\nu)}{\rho} + \frac{1}{\rho} + \frac{z^2}{\rho}\right) + h(r).
$$

To determine function h, we turn to $(5.66)_4$, rewrite it in the form:

$$
u_{r,z} = 2E_{zr} - u_{z,r}\,,
$$

and observe that, for this relation to be consistent with both $(6.26)_4$ and (6.27), function h must have constant value. We take it null. In fact, vector $\boldsymbol{h}_0 = h_0\boldsymbol{e}_1$ would represent an arbitrary translation of the whole space in the vertical direction, the only rigid displacement compatible with the symmetries of the problem and an inevitable indeterminacy, in the absence of Dirichlet boundary conditions, that we lightheartedly dispose of. In conclusion,

$$
\hat{u}^K_z(z,r) = \frac{f}{16\pi G(1-\nu)}\left(\frac{2(1-2\nu)}{\rho} + \frac{1}{\rho} + \frac{z^2}{\rho}\right). \tag{6.28}
$$

References

1. Favata A (2012) On the Kelvin problem. J Elast 109:189–204
2. Love AEH (1927) A treatise on the mathematical theory of elasticity. Cambridge University Press, Cambridge
3. Podio-Guidugli P (2000) A primer in elasticity. Kluwer, Dordrecht
4. Podio-Guidugli P (2004) Examples of concentrated contact interactions in simple bodies. J Elast 75:167–186
5. Sokolnikoff IS (1956) Mathematical theory of elasticity. McGraw-Hill, New York
6. Thompson W (Lord Kelvin) (1848) Note on the integration of the equations of equilibrium of an elastic solid. Cambr Dubl Math J 3:87–89

Part III
Three Other Problems: Melan's, Mindlin's, and Cerruti's

Chapter 7
The Melan and Mindlin Problems

This chapter is devoted to solve the equilibrium problem of a linearly elastic isotropic half-space, subject to a load concentrated at an interior point. The two-dimensional version is named after Ernst Melan (1890–1963), who solved it in 1932 [1]; the three-dimensional version was studied and solved in 1936 [2] by Raymond D. Mindlin (1906–1987), who returned to it some years later [3, 4].

We concentrate of the case of paramount interest in geomechanics, when the load is directed orthogonally to the boundary plane and the Mindlin elastic state is used to compute stresses and soil settlements due to one or more foundation piles. As we shall see, the stress field depends on constitutive choices; no doubt, ordinary soil is far from being elastic and isotropic, and yet Mindlin solution is widely used to estimate footing settlements [5].

7.1 Solution by Superposition

The method we use to solve Melan's and Mindlin's problems is the same, and differs from the methods used by those authors: essentially, as exemplified in the last section of Chap. 1, we proceed by superposition/restriction/super-position.

Our first and main concern is to determine the stress field. This we do in four steps. Preliminarily, we consider a space $\mathcal{S}$ (two-dimensional in Melan's case, three-dimensional in Mindlin's) and we choose an origin $o \in \mathcal{S}$ and a direction $\boldsymbol{e}_1$, so that it makes sense to consider the half-spaces $\mathcal{HS}^{\pm} = \{x \in \mathcal{S} \mid \pm (x - o) \cdot \boldsymbol{e}_1 > 0\}$. Then,

(i) we determine the Kelvin stress $\widetilde{\mathbf{S}}$ induced in $\mathcal{S}$ by a concentrated load $\boldsymbol{f}$ applied at $x = o + a\boldsymbol{e}_1$, $a > 0$;
(ii) we determine the Kelvin stress $\widehat{\mathbf{S}}$ in the same space, this time due to a load $-\boldsymbol{f}$ concentrated at $x = o - a\boldsymbol{e}_1$;

P. Podio-Guidugli and A. Favata, *Elasticity for Geotechnicians*, Solid Mechanics and Its Applications 204, DOI: 10.1007/978-3-319-01258-2_7,

(iii) we consider the restriction $\widetilde{S}$ to $\mathcal{HS}^+$ of the point-wise superposition of the stress fields $\breve{S}$ and $\widehat{S}$, and compute the traction vector $\tilde{s} = -\widetilde{S}e_1$ on the plane boundary of $\mathcal{HS}^+$;

(iv) we superpose to $\widetilde{S}$ a stress field $\overline{S}$ in $\mathcal{HS}^+$ such that the resulting boundary traction $-(\widetilde{S}+\overline{S})e_1$ is null. The stress field solving the M problem at hand is $S^M := \widetilde{S}+\overline{S}$.

We shall go through this sequence of four steps twice, in Sect. 7.2.1 for Melan's problem and in Sect. 7.3.1 for Mindlin's.

Since each of the stress fields we consider is compatible, such is the field S^M. Having found S^M, finding the strain and displacement fields is the matter of routine computations, completely similar to those we made in Chaps. 5 and 6 for the same purposes.

7.2 The Melan Problem

7.2.1 The Stress Field

Preliminarly, we use relations (6.16)–(6.17) to write the components of the stress field for the plane Kelvin problem in a Cartesian frame with the same origin (Fig. 7.1). These components are:

$$
\begin{aligned}
S_{11} &= -\frac{f}{4\pi}\frac{x_1}{(x_1^2+x_2^2)^2}\big((3+\nu_0)x_1^2+(1-\nu_0)x_2^2\big),\\
S_{22} &= \frac{f}{4\pi}\frac{x_1}{(x_1^2+x_2^2)^2}\big((1-\nu_0)x_1^2-(1+3\nu_0)x_2^2\big),\\
S_{12} &= -\frac{f}{4\pi}\frac{x_2}{(x_1^2+x_2^2)^2}\big((3+\nu_0)x_1^2+(1-\nu_0)x_2^2\big).
\end{aligned}
\tag{7.1}
$$

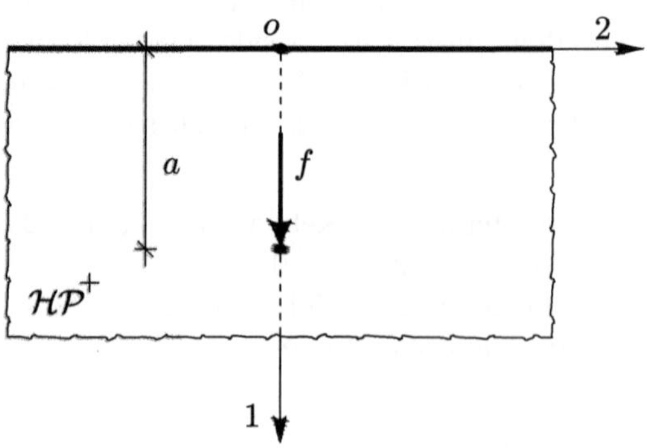

Fig. 7.1 The Melan Problem

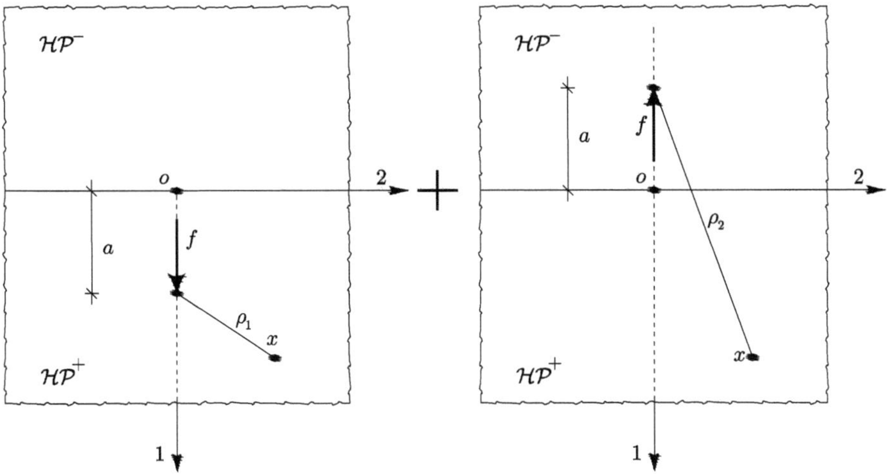

Fig. 7.2 Superposition of mirror-symmetric concentrated loads applied at mirror-symmetric points of half-planes $\mathcal{H}P^+$ and $\mathcal{H}P^-$

7.2.1.1 Steps (i) and (ii)

We use formulas (7.1) twice, to determine the stress fields $\breve{\mathbf{S}}$ and $\widehat{\mathbf{S}}$ induced in $\mathcal{HS}$ by, respectively, a load $\boldsymbol{f} = f\boldsymbol{e}_1$ applied $x = o + a\boldsymbol{e}_1$ and a load $\boldsymbol{f} = -f\boldsymbol{e}_1$ applied $x = o - a\boldsymbol{e}_1$ (Fig. 7.2); all we have to do are two changes in origin. We find:

$$
\begin{aligned}
\breve{S}_{11}(x_1, x_2) &= -\frac{f}{4\pi}\frac{x_1 - a}{((x_1 - a)^2 + x_2^2)^2}\big((3 + \nu_0)(x_1 - a)^2 + (1 - \nu_0)x_2^2\big),\\
\breve{S}_{22}(x_1, x_2) &= \frac{f}{4\pi}\frac{x_1 - a}{((x_1 - a)^2 + x_2^2)^2}\big((1 - \nu_0)(x_1 - a)^2 - (1 + 3\nu_0)x_2^2\big),\\
\breve{S}_{12}(x_1, x_2) &= -\frac{f}{4\pi}\frac{x_2}{((x_1 - a)^2 + x_2^2)^2}\big((3 + \nu_0)(x_1 - a)^2 + (1 - \nu_0)x_2^2\big),
\end{aligned}
\tag{7.2}
$$

and

$$
\begin{aligned}
\widehat{S}_{11}(x_1, x_2) &= \frac{f}{4\pi}\frac{x_1 + a}{((x_1 + a)^2 + x_2^2)^2}\big((3 + \nu_0)(x_1 + a)^2 + (1 - \nu_0)x_2^2\big),\\
\widehat{S}_{22}(x_1, x_2) &= -\frac{f}{4\pi}\frac{x_1 + a}{((x_1 + a)^2 + x_2^2)^2}\big((1 - \nu_0)(x_1 + a)^2 - (1 + 3\nu_0)x_2^2\big),\\
\widehat{S}_{12}(x_1, x_2) &= \frac{f}{4\pi}\frac{x_2}{((x_1 + a)^2 + x_2^2)^2}\big((3 + \nu_0)(x_1 + a)^2 + (1 - \nu_0)x_2^2\big).
\end{aligned}
\tag{7.3}
$$

7.2.1.2 Steps (iii) and (iv)

Component-wise summation of (7.2) and (7.3), followed by restriction to $x_1 \geq 0$, yields the stress field $\widetilde{\mathbf{S}}$ over the closure of $\mathcal{HP}^+$. We quickly see that the traction vector $\tilde{\mathbf{s}} = -\widetilde{\mathbf{S}}(0, x_2)\mathbf{e}_1$ is not null, contrary to Melan's prescription that the traction vector be zero all over the boundary of $\mathcal{HP}^+$. Instead, we have:

$$\begin{aligned} \tilde{\mathbf{s}}(x_2) &= -\widetilde{S}_{11}(0, x_2)\mathbf{e}_1, \\ \widetilde{S}_{11}(0, x_2) &= -\frac{f}{4\pi}\frac{2a\big((3+\nu_0)a^2 + (1-\nu_0)x_2^2\big)}{(a^2+x_2^2)^2}. \end{aligned} \tag{7.4}$$

Therefore, the issue is to find another stress field $\overline{\mathbf{S}}$ over the closure of $\mathcal{HP}^+$, such that

$$(\widetilde{\mathbf{S}}(0, x_2) + \overline{\mathbf{S}}(0, x_2))\mathbf{e}_1 \equiv \mathbf{0}.$$

We construct $\overline{\mathbf{S}}$ by using the Boussinesq-Flamant stress field as a *stress Green function* (a notion we introduced in the simpler context of Sect. 1.2).

The Cartesian components of the Boussinesq-Flamant plane stress field can be easily deduced from (4.14); they are:

$$\begin{aligned} S_{11}^{BF}(x_1, x_2) &= -\frac{2f}{\pi}\frac{x_1^3}{(x_1^2+x_2^2)^2}, \\ S_{22}^{BF}(x_1, x_2) &= -\frac{2f}{\pi}\frac{x_1 x_2^2}{(x_1^2+x_2^2)^2} \\ S_{12}^{BF}(x_1, x_2) &= -\frac{2f}{\pi}\frac{x_1^2 x_2}{(x_1^2+x_2^2)^2}. \end{aligned}$$

The components of the *Green tensor* $\mathbf{G}^{BF}$ are obtained from those of $\mathbf{S}^{BF}$ by setting $f = 1$ and replacing x_2 by $(x_2 - \xi)$, that is, relocating the origin on the plane $x_1 = 0$ (see Fig. 7.3). These measures yield:

$$\begin{aligned} \widehat{G}_{11}^{BF}(x_1, x_2; \xi) &= -\frac{2}{\pi}\frac{x_1^3}{(x_1^2+(x_2-\xi)^2)^2}, \\ \widehat{G}_{22}^{BF}(x_1, x_2; \xi) &= -\frac{2}{\pi}\frac{x_1 x_2^2}{(x_1^2+(x_2-\xi)^2)^2}, \\ \widehat{G}_{12}^{BF}(x_1, x_2; \xi) &= -\frac{2}{\pi}\frac{x_1^2 x_2}{(x_1^2+(x_2-\xi)^2)^2}. \end{aligned}$$

We are now in position to determine the tensor $\overline{\mathbf{S}}$:

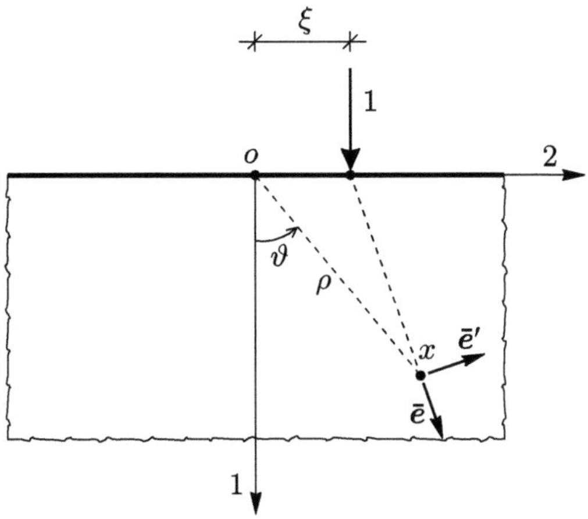

Fig. 7.3 The origin relocation that permits to deduce $\boldsymbol{G}^{BF}$ from $\boldsymbol{S}^{BF}$

$$\overline{\mathbf{S}}(x_1, x_2) = \int_{-\infty}^{+\infty} \hat{p}(x_1, \xi)\widehat{\boldsymbol{G}}^{BF}(x_1, x_2; \xi)\, d\xi,$$

where the load function $\hat{p}$ is the negative of the surface traction (7.4) that we want to eliminate:

$$\hat{p}(x_1, \xi) := \frac{f}{4\pi} \frac{2a}{(a^2 + \xi^2)^2} \left((3 + \nu_0)a^2 + (1 - \nu_0)\xi^2\right).$$

Finding $\overline{\mathbf{S}}$ is the matter of a nontrivial computation, whose development is the same for all components; we here sketch it for the first component, details are found in Appendix A.7.

To begin with, we have that

$$\begin{aligned}\overline{S}_{11}(x_1, x_2) &= \int_{-\infty}^{+\infty} \hat{p}(x_1, \xi)\widehat{G}_{11}(x_1, x_2; \xi)\, d\xi \\ &= -\frac{afx_1^3}{\pi^2}\Big(a^2(3 + \nu_0)I_1 + (1 - \nu_0)I_2\Big),\end{aligned} \tag{7.5}$$

with

$$I_1 := \int_{-\infty}^{+\infty} \frac{1}{(a^2 + \xi^2)^2(x_1^2 + (x_2 - \xi)^2)^2},$$
$$I_2 := \int_{-\infty}^{+\infty} \frac{\xi^2}{(a^2 + \xi^2)^2(x_1^2 + (x_2 - \xi)^2)^2}.$$

A lengthy computation based on the methods of residues yields:

$$I_1 = \frac{2\pi}{4a^3x_1^3}\frac{(x_1+a)^3(x_1^2+3ax_1+a^2)+(a^3+x_1^3)x_2^2}{\big((x_1+a)^2+x_2^2\big)^3},$$
$$I_2 = \frac{\pi}{2ax_1^3}\frac{x_1^2(x_1+a)^3+(x_1+a)(x_1^2+5ax_1+a^2)x_2^2+ax_2^2}{\big((x_1+a)^2+x_2^2\big)^3}. \tag{7.6}$$

Substituting (7.6) into (7.5) we arrive at:

$$\begin{aligned}\overline{S}_{11}(x_1,x_2) = &-\frac{f}{2\pi\big((x_1+a)^2+x_2^2\big)^3}\Big((3+\nu_0)\big((x_1+a)^3(x_1^2+3ax_1+a^2)\\ &+(a^3+x_1^3)x_2^2\big)+(1-\nu_0)\big(x_1^2(x_1+a)^3\\ &+(x_1+a)(x_1^2+5ax_1+a^2)x_2^2+ax_2^4\big)\Big).\end{aligned}$$

With this, we are ready to write the first component of the Melan stress tensor:

$$\begin{aligned}\widehat{S}^{Me}_{11}(x_1,x_1) &= \widetilde{S}_{11}+\overline{S}_{11}\\ &= -\frac{f}{2\pi}\Bigg((1+\nu_0)\bigg(\frac{(x_1-a)^3}{\rho_1^4}+\frac{(x_1+a)\big((x_1+a)^2+2ax_1\big)}{\rho_2^4}-\frac{8ax_1(a+x_1)x_2^2}{\rho_2^6}\bigg)\\ &\quad+\frac{1-\nu_0}{2}\bigg(\frac{x_1-a}{\rho_1^2}+\frac{3x_1+a}{\rho_2^2}-\frac{4x_1x_2^2}{\rho_2^4}\bigg)\Bigg),\end{aligned}$$

where

$$\rho_1 := \sqrt{(x_1-a)^2+x_2^2}, \quad \rho_2 := \sqrt{(x_1+a)^2+x_2^2}.$$

The other two components are found in a completely analogous manner. Their expressions are:

$$\begin{aligned}(2\pi f^{-1})\widehat{S}^{Me}_{22}(x_1,x_2) = &-\Bigg((1+\nu_0)\bigg(\frac{(x_1-a)x_2^2}{\rho_1^4}+\frac{(x_1+a)(x_2^2+2a^2)-2ax_2^2}{\rho_2^4}\\ &+\frac{8ax_1(a+x_1)x_2^2}{\rho_2^6}\bigg)+\frac{1-\nu_0}{2}\bigg(-\frac{x_1-a}{\rho_1^2}+\frac{x_1+3a}{\rho_2^2}+\frac{4x_1x_2^2}{\rho_2^4}\bigg)\Bigg),\end{aligned}$$

$$\begin{aligned}(2\pi f^{-1})\widehat{S}^{Me}_{12}(x_1,x_2) = &-x_2\Bigg((1+\nu_0)\bigg(\frac{(x_1-a)^2}{\rho_1^4}+\frac{x_1^2-2ax_1-a^2}{\rho_2^4}\\ &+\frac{8ax_1(a+x_1)^2}{\rho_2^6}\bigg)+\frac{1-\nu_0}{2}\bigg(\frac{1}{\rho_1^2}-\frac{1}{\rho_2^2}-\frac{4x_1(a+x_1)}{\rho_2^4}\bigg)\Bigg).\end{aligned}$$

7.2.2 The Strain and Displacement Fields

As we have done systematically so far, we obtain the Melan strain field by inserting the stress field we just obtained in the inverse constitutive law (2.57). After some manipulations, we have:

$$
\begin{aligned}
&(2\pi E_0 f^{-1})\,\widehat{E}^{Me}_{11}(x_1,x_2)\\
&\quad= -(1+\nu_0)\Bigg(\frac{(x_1-a)^3-\nu_0(x_1-a)x_2^2}{\rho_1^4}\\
&\qquad+\frac{(x_1+a)\big((x_1+a)^2+2ax_1\big)-\nu_0\big((x_1+a)(x_2^2+2a^2)-2ax_2^2\big)}{\rho_2^4}\\
&\qquad-\frac{(1+\nu_0)8ax_1(a+x_1)x_2^2}{\rho_2^6}\Bigg)+\frac{1-\nu_0}{2}\Bigg(\frac{(1+\nu_0)(x_1-a)}{\rho_1^2}\\
&\qquad+\frac{3x_1+a-\nu_0(x_1+3a)}{\rho_2^2}-\frac{(1-\nu_0)4x_1x_2^2}{\rho_2^4}\Bigg),
\end{aligned}
\tag{7.7}
$$

$$
\begin{aligned}
&(2\pi E_0 f^{-1})\,\widehat{E}^{Me}_{22}(x_1,x_2)\\
&\quad= -(1+\nu_0)\Bigg(\frac{(x_1-a)x_2^2-\nu_0(x_1-a)^3}{\rho_1^4}\\
&\qquad+\frac{(x_1+a)(x_2^2+2a^2)-2ax_2^2-\nu_0(x_1+a)\big((x_1+a)^2+2ax_1\big)}{\rho_2^4}\\
&\qquad(1+\nu_0)\frac{8ax_1(a+x_1)x_2^2}{\rho_2^6}\Bigg)+\frac{1-\nu_0}{2}\Bigg(-(1+\nu_0)\frac{x_1-a}{\rho_1^2}\\
&\qquad+\frac{x_1+3a-\nu_0(3x_1+a)}{\rho_2^2}-(1-\nu_0)\frac{4x_1x_2^2}{\rho_2^4}\Bigg)\Bigg),
\end{aligned}
\tag{7.8}
$$

$$
\begin{aligned}
(2\pi E_0 f^{-1})\,\widehat{E}^{Me}_{12}(x_1,x_2) &= -(1+\nu_0)x_2\Bigg((1+\nu_0)\Bigg(\frac{(x_1-a)^2}{\rho_1^4}+\frac{x_1^2-2ax_1-a^2}{\rho_2^4}\\
&\quad+\frac{8ax_1(a+x_1)^2}{\rho_2^6}\Bigg)+\frac{1-\nu_0}{2}\Bigg(\frac{1}{\rho_1^2}-\frac{1}{\rho_2^2}-\frac{4x_1(a+x_1)}{\rho_2^4}\Bigg)\Bigg).
\end{aligned}
$$

In order to determine the displacement field, we have to solve the following system of PDEs:

$$
u_{1,1}=E^{Me}_{11},\quad u_{2,2}=E^{Me}_{22},\quad u_{1,2}+u_{2,1}=2\,E^{Me}_{12},
\tag{7.9}
$$

subject to the symmetry conditions:

$$
\hat{u}_1(x_1,x_2)=\hat{u}_1(x_1,-x_2),\quad \hat{u}_2(x_1,x_2)=-\hat{u}_2(x_1,-x_2).
\tag{7.10}
$$

With the use of (7.7) and (7.8), integration of 7.9_1 and 7.9_2 yields:

$$\begin{aligned}\hat{u}_1^{Me}(x_1,x_2) &= \int_{\bar{x}_1}^{x_1} \widehat{E}_{11}^{Me}(s,x_2)\,ds + \hat{g}_1(x_2)\\ &= -\frac{f}{8\pi E_0}\Bigg(\frac{2(1+\nu_0)x_2^2}{\rho_1^2} - \frac{2(1+\nu_0)\big(2ax_1(1+\nu_0)-(3-\nu_0)x_2^2\big)}{\rho_2^2}\\ &\quad + \frac{8a(1+\nu_0)^2x_1x_2^2}{\rho_2^4} + (3-\nu_0)(1+\nu_0)\log\rho_1\\ &\quad + (5-(2-\nu_0)\nu_0)\log\rho_2\Bigg) + \hat{g}_1(x_2),\end{aligned} \tag{7.11}$$

$$\begin{aligned}\hat{u}_2^{Me}(x_1,x_2) &= \int_{\bar{x}_2}^{x_2} \widehat{E}_{22}^{Me}(x_1,s)\,ds + \hat{g}_2(x_1)\\ &= \frac{f}{4\pi E_0}\Bigg((1+\nu_0)(x_1-a)x_2\left(\frac{1+\nu_0}{\rho_1^2}+\frac{3-\nu_0}{\rho_2^2}\right)\\ &\quad + \frac{4a(1+\nu_0)^2x_1x_2(x_1+a)}{\rho_2^4} - 4(1-\nu_0)\arctan\left(\frac{x_2}{x_1+a}\right)\Bigg) + \hat{g}_2(x_1).\end{aligned} \tag{7.12}$$

Note that the symmetry condition $(7.10)_2$ implies that

$$\hat{g}_2(x_1) \equiv 0. \tag{7.13}$$

To determine function $\hat{g}_1$, we insert in $(7.9)_3$ relations (7.11) and (7.12) (with (7.13) taken into account), and find out that:

$$\Big(\int_{\bar{x}_1}^{x_1} \widehat{E}_{11}^{Me}(s,x_2)\,ds\Big)_{,2} + \Big(\int_{\bar{x}_2}^{x_2} \widehat{E}_{22}^{Me}(x_1,s)\,ds\Big)_{,1} - 2\,\widehat{E}_{12}^{Me}(x_1,x_2) = 0 = \hat{g}_1'(x_2).$$

Hence, function $\hat{g}_1$ must be constant-valued; we dispose of the residual irrelevant indeterminacy by taking the relative constant null.

7.3 The Mindlin Problem

To solve this problem (Fig. 7.4), we take once more the four steps listed sequentially in Sect. 7.1, this time with considerable analytical complications.

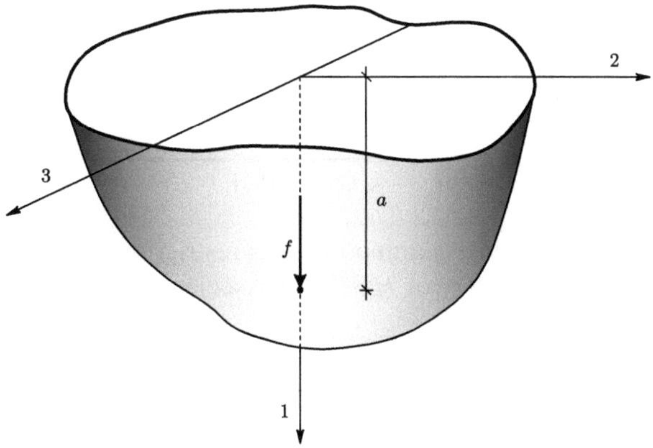

Fig. 7.4 The Mindlin Problem

7.3.1 The Stress Field

7.3.1.1 Steps (i) and (ii)

To determine the fields $\check{S}$ and $\widehat{S}$ over the whole space $\mathcal{S}$, we make use of the solution of the 3-D Kelvin Problem. From (6.25), with two appropriate changes in origin, we deduce that

$$\check{S}_{zz}(z,r) = -\frac{f}{8\pi(1-\nu)}\left(3\frac{(z-a)^3}{\rho_1^5} - (1-2\nu)\frac{z-a}{\rho_1^3}\right),$$

$$\check{S}_{rr}(z,r) = -\frac{f}{8\pi(1-\nu)}\left(3\frac{(z-a)r^2}{\rho_1^5} + (1-2\nu)\frac{z-a}{\rho_1^3}\right),$$

$$\check{S}_{\varphi\varphi}(z,r) = \frac{f(1-2\nu)}{8\pi(1-\nu)}\frac{z-a}{\rho_1^3},$$

$$\check{S}_{zr}(z,r) = -\frac{f}{8\pi(1-\nu)}\left(3\frac{(z-a)^2 r}{\rho_1^5} - (1-2\nu)\frac{z-a}{\rho_1^3}\right),$$

and

$$\widehat{S}_{zz}(z,r) = \frac{f}{8\pi(1-\nu)}\left(3\frac{(z+a)^3}{\rho_1^5} - (1-2\nu)\frac{z+a}{\rho_2^3}\right),$$

$$\widehat{S}_{rr}(z,r) = \frac{f}{8\pi(1-\nu)}\left(3\frac{(z+a)r^2}{\rho_2^5} + (1-2\nu)\frac{z+a}{\rho_2^3}\right),$$

$$\widehat{S}_{\varphi\varphi}(z,r) = -\frac{f(1-2\nu)}{8\pi(1-\nu)}\frac{z+a}{\rho_2^3},$$

$$\widehat{S}_{zr}(z,r) = \frac{f}{8\pi(1-\nu)}\left(3\frac{(z+a)^2 r}{\rho_2^5} - (1-2\nu)\frac{z+a}{\rho_2^3}\right),$$

where

$$\rho_1 := \sqrt{(z-a)^2+r^2}, \qquad \rho_2 := \sqrt{(z+a)^2+r^2}. \tag{7.14}$$

Component-wise summation plus restriction to the half-space $\mathcal{HS}^+$ yield the field $\widetilde{\mathbf{S}}$; at the boundary of $\mathcal{HS}^+$, the associated traction vector is:

$$\begin{aligned}\tilde{\mathbf{s}} &= -\widetilde{S}_{11}(0,x_2,x_3)\mathbf{e}_1,\\ \widetilde{S}_{11}(0,x_2,x_3) &= -\frac{f}{4\pi}\frac{a\big(2a^2(2-\nu)+(x_2^2+x_3^2)(1-2\nu)\big)}{(1-\nu)\big(a^2+x_2^2+x_3^2\big)^{\frac{5}{2}}}\end{aligned} \tag{7.15}$$

(cf. (7.4)).

7.3.1.2 Steps (iii) and (iv)

To eliminate the effect of the undesired surface traction (7.15), we have to superimpose to $\widetilde{\mathbf{S}}$ a stress field

$$\overline{\mathbf{S}}(x_1,x_2,x_3) = \int_{-\infty}^{+\infty}\int_{-\infty}^{+\infty} \hat{p}(\eta,\zeta)\widehat{\mathbf{G}}^B(x_1,x_1,x_3;\eta,\zeta)\,d\eta d\zeta,$$

where $\mathbf{G}^B$ is the stress Green function associated with the Boussinesq stress $\mathbf{S}^B$, and where

$$\hat{p}(\eta,\zeta) := \frac{f}{4\pi}\frac{a\big(2a^2(2-\nu)+(\eta^2+\zeta^2)(1-2\nu)\big)}{(1-\nu)\big(a^2+\eta^2+\zeta^2\big)^{\frac{5}{2}}}, \tag{7.16}$$

(cf. (7.15)). Hereafter, we exemplify the construction of $\overline{\mathbf{S}}$, a cumbersome task indeed, by undertaking it for the component $\overline{S}_{11}$.

The integral in question is:

$$\overline{S}_{11}(x_1,x_2,x_3) = \int_{-\infty}^{+\infty}\int_{-\infty}^{+\infty} \hat{p}(\eta,\zeta)\,\widehat{G}_{11}^B(x_1,x_1,x_3;\eta,\zeta)\,d\eta d\zeta. \tag{7.17}$$

where, in view of $(5.62)_1$,

$$\widehat{G}_{11}^B(x_1,x_1,x_3;\eta,\zeta) = -\frac{3}{2\pi}\frac{x_1^3}{\big(x_1^2+(x_2-\eta)^2+(x_3-\zeta)^2\big)^{\frac{5}{2}}}.$$

We have been unable to come up with an explicit evaluation for long, until we found the following circuitous route.[1]

We recall from Sect. 5.8 that the Boussinesq stress field can be given the Boussinesq-Papkovitch-Neuber representation (A.16) in terms of two harmonic functions ψ^B and φ^B. In particular, component S^B_{11} admits the representation (5.77), that we here recall for the reader's convenience:

$$S^B_{11} = \frac{1}{1-\nu}\big(2(1-\nu)\psi^B_{,1} - \varphi^B_{,11} - x_1\psi^B_{,11}\big),$$

where

$$\psi^B = \frac{1}{2\pi\rho}, \quad \varphi^B = \frac{1-2\nu}{2\pi}\log(x_1+\rho), \quad \rho^2 := x_1^2 + x_2^2 + x_3^2.$$

Accordingly, the associated stress Green function G^B_{11} turns out to be:

$$G^B_{11} = \frac{1}{1-\nu}\big(2(1-\nu)\gamma_{1,1} - \gamma_{2,11} - x_1\gamma_{1,11}\big),$$

where

$$\gamma_1 = \widehat{\gamma}_1(x_1, x_2, x_3; \eta, \zeta) := \widehat{\psi}^B(x_1, x_2-\eta, x_3-\zeta),$$
$$\gamma_2 = \widehat{\gamma}_2(x_1, x_2, x_3; \eta, \zeta) := \widehat{\varphi}^B(x_1, x_2-\eta, x_3-\zeta).$$

And, the stress component $\overline{S}_{11}$ we are looking for can be given the following form:

$$\overline{S}_{11} = \frac{1}{1-\nu}\big(2(1-\nu)\overline{\psi}_{,1} - \overline{\varphi}_{,11} - x_1\overline{\psi}_{,11}\big),$$

where the harmonic functions $\overline{\psi}$ and $\overline{\varphi}$ have the following expressions in terms of the harmonic functions γ_1 and γ_2:

$$\overline{\psi} = \int_{\partial\mathcal{H}^+} \hat{p}(\eta,\zeta)\,\widehat{\gamma}_1(x_1, x_1, x_3; \eta, \zeta)\, d\eta d\zeta,$$
$$\overline{\varphi} = \int_{\partial\mathcal{H}^+} \hat{p}(\eta,\zeta)\,\widehat{\gamma}_2(x_1, x_1, x_3; \eta, \zeta)\, d\eta d\zeta. \tag{7.18}$$

Thus, in place of the awkward integral (7.17), our task is to compute the integrals (7.18). This is doable, with the use of certain well-known properties of harmonic functions.

---

[1] We are indebted to Professor G. Tarantello for many useful conversations on the matters; our techniques are akin to those used in [4] and [6].

To begin with, recall (from [7], say) *Green's second identity*:

$$\int_{\Omega} u\Delta v - v\Delta u = \int_{\partial\Omega} u\frac{\partial v}{\partial \mathbf{n}} - v\frac{\partial u}{\partial \mathbf{n}}, \tag{7.19}$$

where u and v are scalar fields defined over region Ω, whose boundary $\partial\Omega$ has an a.e. well-defined outward normal $\mathbf{n}$. We apply this identity for $\Omega \equiv \mathcal{HS}^+$, u a harmonic function, and v the solution Γ of the boundary-value problem:

$$\begin{cases} \Delta\widehat{\Gamma}(x_1, x_2, x_3) = \delta(x_1 + a, x_2, x_3) & \text{in } \mathcal{HS}^+, \\ \widehat{\Gamma}(x_1, x_1, x_3) = 0 & \text{on } \partial\mathcal{HS}^+, \end{cases} \tag{7.20}$$

where $\delta(x_1 + a, x_2, x_3)$ is the Dirac delta function (see Sect. A.1) centered at point $x = o + a\boldsymbol{e}_1$, namely,

$$\Gamma = \widehat{\Gamma}(x_1, x_2, x_3) := \frac{1}{4\pi}\left(\frac{1}{\sqrt{(x_1 - a)^2 + x_2^2 + x_3^2}} - \frac{1}{\sqrt{(x_1 + a)^2 + x_2^2 + x_3^2}}\right).$$

we find:

$$\widehat{u}(x_1 + a, x_2, x_3) = \int_{\partial\mathcal{HS}^+} \widehat{u}(x_1, \eta, \zeta)\frac{\partial\widehat{\Gamma}}{\partial x_1}(x_1, \eta, \zeta). \tag{7.21}$$

Moreover, function $\hat{p}$ in (7.16) can be written as follows in terms of the normal derivative of Γ:

$$\hat{p}(\eta, \zeta) = \frac{f}{2(1-\nu)}\left(2(1-\nu)\frac{\partial\widehat{\Gamma}}{\partial x_1}(x_1, x_2, x_3) - a\frac{\partial^2\widehat{\Gamma}}{\partial a\partial x_1}(x_1, x_2, x_3)\right)\Big|_{(0,\eta,\zeta)}.$$

With this, integrals (7.18) take the convenient form:

$$\overline{\psi}(x_1, x_2, x_3) = -\frac{f}{2(1-\nu)}\left(2(1-\nu)\int_{\partial\mathcal{H}^+} \widehat{\gamma}_1\frac{\partial\widehat{\Gamma}}{\partial x_1} - a\frac{\partial}{\partial a}\int_{\partial\mathcal{H}^+} \widehat{\gamma}_1\frac{\partial\widehat{\Gamma}}{\partial x_1}\right) \tag{7.22}$$

and

$$\overline{\varphi}(x_1, x_2, x_3) = -\frac{f}{8(1-\nu)}\left(2(1-\nu)\int_{\partial\mathcal{H}^+} \widehat{\gamma}_2\frac{\partial\widehat{\Gamma}}{\partial x_1} - a\frac{\partial}{\partial a}\int_{\partial\mathcal{H}^+} \widehat{\gamma}_2\frac{\partial\widehat{\Gamma}}{\partial x_1}\right). \tag{7.23}$$

To evaluate the integrals in the right sides of (7.22) and (7.23), we make use of (7.21) and find:

$$\int_{\partial\mathcal{H}^+} \widehat{\gamma}_1 \frac{\partial \widehat{\Gamma}}{\partial x_1} = \widehat{\psi}^B(x_1 + a, x_2, x_3),$$

$$\int_{\partial\mathcal{H}^+} \widehat{\gamma}_2 \frac{\partial \widehat{\Gamma}}{\partial x_1} = \widehat{\varphi}^B(x_1 + a, x_2, x_3),$$

whence

$$\overline{\psi} = -\frac{f}{2\pi}\left(\frac{a(x_1+a)}{2\rho_2^3} + \frac{1-\nu}{\rho_2}\right),$$

$$\overline{\varphi} = -\frac{f}{2\pi}\left(\frac{a(x_1+a)}{2\rho_2^3} + (1-2\nu)\log(x_1 + a + \rho_2)\right).$$

All in all, the first component of the stress tensor field solving the Mindlin Problem is:

$$S_{11}^{Mi} = \frac{f}{8\pi(1-\nu)}\left(-\frac{(1-2\nu)(x_1-a)}{\rho_1^3} + \frac{(1-2\nu)(x_1-a)}{\rho_2^3} - \frac{3(x_1-a)^2}{\rho_1^5} - \frac{3(3-4\nu)x_1(x_1+a)^2 - 3a(x_1+a)(5x_1-a)}{\rho_2^5} - \frac{30ax_1(x_1+a)^3}{\rho_2^7}\right) \tag{7.24}$$

(ρ_1 and ρ_2 are defined in (7.14)).

At the expenses of completely similar long computations, the remaining stress components are found to be:

$$S_{rr}^{Mi} = \frac{f}{8\pi(1-\nu)}\left(\frac{(1-2\nu)(z-a)}{\rho_1^3} - \frac{(1-2\nu)(z+7a)}{\rho_2^3} + \frac{4(1-\nu)(1-2\nu)}{\rho_2(\rho_2+z+a)} - \frac{3r^2(z-a)}{\rho_1^5} + \frac{6a(1-2\nu)(z+a)^2 - 6a^2(z+a) - 3(3-4\nu)r^2(z-a)}{\rho_2^5} - \frac{30ar^2z(z+a)}{\rho_2^7}\right),$$

$$S_{\varphi\varphi}^{Mi} = \frac{f(1-2\nu)}{8\pi(1-\nu)}\left(\frac{(z-a)}{\rho_1^3} + \frac{(3-4\nu)(z+a) - 6a}{\rho_2^3} - \frac{4(1-\nu)}{\rho_2(\rho_2+z+a)} + \frac{6a(z+a)^2}{\rho_2^5} - \frac{6a^2(z+a)}{(1-2\nu)\rho_2^5}\right),$$

$$S_{zr}^{Mi} = \frac{fr}{8\pi(1-\nu)}\left(-\frac{1-2\nu}{\rho_1^3} + \frac{1-2\nu}{\rho_2^3} - \frac{3(z-a)^2}{\rho_1^5} - \frac{30az(z+a)^2}{\rho_2^7}\right.$$
$$\left. - \frac{3(3-4\nu)z(z+a) - 3a(3z+a)}{\rho_2^5}\right). \tag{7.25}$$

Note that, the Boussinesq stress (5.62) is recovered for $a = 0$.

7.3.2 The Strain and Displacement Fields

The Mindlin displacement strain field is found by insertion of the stress representations (7.24) and (7.25) into the inverse constitutive equation (2.45). After some algebraic manipulations, one finds:

$$(-16\pi G(1-\nu)(1+\nu)f^{-1})E_{zz}^{Mi}$$
$$= \frac{3(z-a)\left(a^2 - 2az - \nu r^2 + z^2\right)}{\rho_1^5}$$
$$+ \left(\frac{30az(a+z)\left(a^2+2az-\nu r^2+z^2\right)}{\rho_2^7} - \frac{3}{\rho_2^5}\left(a^3\left(4\nu^2-1\right) + a^2\left(12\nu^2 z + z\right)\right.\right.$$
$$\left. + a\left(\nu(4\nu-3)r^2 + \left(8\nu^2+4\nu-1\right)z^2\right) + (4\nu-3)z\left(z^2 - \nu r^2\right)\right)$$
$$\left. + (2\nu-1)\frac{\left(a\left(4\nu^2+10\nu-1\right) + \left(4\nu^2-2\nu+1\right)z\right)}{\rho_2^3}\right) - \frac{\left(4\nu^2-1\right)(z-a)}{\rho_1^3},$$

$$(16\pi(1-\nu)Gf^{-1})E_{rr}^{Mi}$$
$$= \frac{-6a^2(a+z) - 3(4\nu-3)r^2(a-z) + 6a(1-2\nu)(a+z)^2}{\rho_2^5}$$
$$+ \frac{3r^2(a-z)}{\rho_1^5} - \frac{30ar^2z(a+z)}{\rho_2^7} + \frac{4(\nu-1)(2\nu-1)}{\rho_2(a+\rho_2+z)} + \frac{(2\nu-1)(a-z)}{\rho_1^3}$$
$$+ \frac{(2\nu-1)(7a+z)}{\rho_2^3} - \frac{\nu\left(3(a-z)\left(a^2-2az+r^2+z^2\right)\right)}{(\nu+1)\rho_1^5}$$
$$+ \frac{\nu\left(30az(a+z)\left(a^2+2az+r^2+z^2\right)\right)}{(\nu+1)\rho_2^7}$$
$$+ \frac{\nu\left(3\left(a^3(4\nu+1)+a^2(8\nu-5)z+a\left((4\nu-3)r^2-3z^2\right)-(4\nu-3)z\left(r^2+z^2\right)\right)\right)}{(\nu+1)\rho_2^5}$$
$$- \frac{\nu\left((2\nu-1)(a(4\nu+11)+(4\nu-3)z)\right)}{(\nu+1)\rho_2^3} + \frac{(2\nu-1)(a-z)}{(\nu+1)\rho_1^3},$$

$$
\begin{aligned}
&(16\pi(1-\nu)Gf^{-1})E^{Mi}_{\varphi\varphi}\\
&\quad=(1-2\nu)\left(-\frac{6a^2(a+z)}{\rho_2^5}+\frac{4(\nu-1)}{\rho_2\,(a+\rho_2+z)}\right.\\
&\qquad\left.+\frac{(3-4\nu)(a+z)-6a}{\rho_2^3}+\frac{6a(a+z)^2}{\rho_2^5}+\frac{z-a}{\rho_1^3}\right)\\
&\qquad-\frac{3\nu(a-z)\left(a^2-2az+r^2+z^2\right)}{(\nu+1)\rho_1^5}\\
&\qquad+\frac{30\nu az(a+z)\left(a^2+2az+r^2+z^2\right)}{(\nu+1)\rho_2^7}\\
&\qquad+\frac{3\nu\left(a^3(4\nu+1)+a^2(8\nu-5)z+a\left((4\nu-3)r^2-3z^2\right)-(4\nu-3)z\left(r^2+z^2\right)\right)}{(\nu+1)\rho_2^5}\\
&\qquad-\frac{3\nu(2\nu-1)(a(4\nu+11)+(4\nu-3)z)}{(\nu+1)\rho_2^3}-\frac{\nu(2\nu-1)(a-z)}{(\nu+1)\rho_1^3},
\end{aligned}
$$

and

$$
\begin{aligned}
&(16\pi G(1-\nu)(1+\nu)f^{-1})E^{Mi}_{zr}\\
&\quad=-3\frac{(a-z)\left(a^2\nu+a(\nu r+r-2\nu z)+\nu r^2-(\nu+1)rz+\nu z^2\right)}{\rho_1^5}\\
&\qquad+\frac{3\left(a^3\nu(4\nu+1)+a^2((\nu+1)r+\nu(8\nu-5)z)+a\nu\left((4\nu-3)r^2+4(\nu+1)rz-3z^2\right)\right)}{\rho_2^5}\\
&\qquad+\frac{3\left(-(4\nu-3)z\left(\nu r^2-(\nu+1)rz+\nu z^2\right)\right)}{\rho_2^5}\\
&\qquad+\frac{30az(a+z)\left(a^2\nu-a(\nu r+r-2\nu z)+\nu r^2-(\nu+1)rz+\nu z^2\right)}{\rho_2^7}\\
&\qquad-\frac{(2\nu-1)(\nu(a(4\nu+11)+(4\nu-3)z)+(\nu+1)r)}{\rho_2^3}\\
&\qquad+\frac{(2\nu-1)(-a\nu+\nu r+r+\nu z)}{\rho_1^3}.
\end{aligned}
$$

Note that, if $a=0$, the deformation field reduces to Boussinesq's, as given by (5.65).

To find the displacement field, we could follow a by now familiar course, and exploit the compatibility equation (2.9) as we did in SubSect. 7.2.2. However, we prefer to perform this task by employing the same procedure we adopted for the stress field, namely,

(i) we superimpose the Kelvin displacements $\breve{\boldsymbol{u}}$ and $\widehat{\boldsymbol{u}}$ corresponding to the stress fields $\breve{\boldsymbol{S}}$ and $\widehat{\boldsymbol{S}}$ defined in Sect. 7.3.1;

(ii) we consider the restriction $\tilde{\boldsymbol{u}}$ to $\mathcal{HS}^+$ of the above point-wise superposition, and we further superimpose to it the displacement field $\overline{\boldsymbol{u}}$, that we determine by means of (7.18) and (5.76).

The outcome is the Mindlin displacement field, in cylindrical coordinates:

$$u_z^{Mi} = \frac{fr}{16\pi G(1-\nu)}\left(\frac{z-a}{\rho_1^3} + \frac{(3-4\nu)(z-a)}{\rho_2^3} - \frac{4(1-\nu)(1-2\nu)}{\rho_2(\rho_2+z+a)} + \frac{6az(z+a)}{\rho_2^5}\right),$$

$$u_r^{Mi} = \frac{f}{16\pi G(1-\nu)}\left(\frac{3-4\nu}{\rho_1} + \frac{8(1-\nu)^2-(3-4\nu)}{\rho_2} + \frac{(z-a)^2}{\rho_1^3} + \frac{(3-4\nu)(z+a)^2-2az}{\rho_2^3} + \frac{6az(z+a)^2}{\rho_2^5}\right).$$

References

1. Melan E (1932) Der Spannungszustand der durch eine Einzelkraft im Innern beanspruchten Halbscheibe. Z Angew Math Mech 12:343–346
2. Mindlin RD (1936) Force at a point in the interior of a semi-infinite solid. Physics 7:195–202
3. Mindlin RD (1950) Nuclei of strain in the semi-infinite solid. J Appl Phys 21:926–930
4. Mindlin RD (1953) Force at a point in the interior of a semi-infinite solid. In: Proceedings of 1st Midwest conference solid mechanics, pp 56–69
5. Bowles JE (1996) Foundation Analysis and Design. Mcgraw-Hill, New York
6. Storåkers B (1985) A note on superposition in Mindlin's problem. J Appl Math Pys 36:927–932
7. Evans LC (2010) Partial Differential Equations. American Mathematical Society, Providence

Chapter 8
The Cerruti Problem

The equilibrium problem solved by Valentino Cerruti (1850–1909) concerns a linearly elastic isotropic half-space acted upon by a *concentrated* load, *tangent* to the boundary plane [1].

Recall that the load is *perpendicular* to the boundary plane in both Boussinesq's and Flamant's problems, concentrated in the former and diffused over a line in the latter. Load tangency, no matter if concentrated or diffused, brings about a different type of symmetries. We exemplify them by taking up a version of the original Cerruti Problem, where a *diffused* tangent load $\boldsymbol{f} = f\boldsymbol{e}_2$ is applied, with constant magnitude per unit length and infinitely long support (Fig. 8.1).

We solve this problem by adopting the same strategy as for the Flamant Problem (Chap. 4), that is, by exploiting its intrinsic symmetries to determine, in the first place, a divergenceless plane stress field balancing the applied load and compatible with the construction to follow of a plane displacement field, having the forecasted symmetries.

8.1 Displacement and Stress Symmetries

For the diffused-load version of the Cerruti Problem we here study, intuition suggests that, in the closure of the half-space $\mathcal{HS}^+$, *the displacement field must be plane and independent of coordinate* x_3:

$$\boldsymbol{u} = \widehat{u}_\alpha(\rho, \vartheta)\boldsymbol{e}_\alpha,$$

an anticipation that can be argued just as we did in Sect. 4.1 for the Flamant Problem. However, the parities induced by a tangent applied load are not the same as when the load is perpendicular to the plane boundary of $\mathcal{HS}^+$.

A short reflection suggests the following prediction:

$$\widehat{u}_1(\rho, \vartheta) = -\widehat{u}_1(\rho, -\vartheta), \quad \widehat{u}_2(\rho, \vartheta) = \widehat{u}_2(\rho, -\vartheta) \tag{8.1}$$

P. Podio-Guidugli and A. Favata, *Elasticity for Geotechnicians*, Solid Mechanics and Its Applications 204, DOI: 10.1007/978-3-319-01258-2_8,

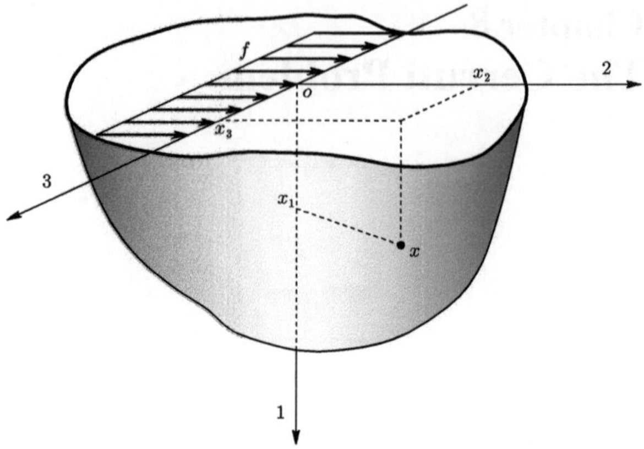

Fig. 8.1 The Cerruti Problem

(Fig. 8.2). Here is how we argue this. At a a typical point of coordinates (ρ, ϑ), let us agree that the displacement components have the direction shown; accordingly, were the load reversed, the displacement components at the mirror point of coordinates $(\rho, -\vartheta)$ would be:

$$\widehat{u}_1(\rho, -\vartheta; -f) = \widehat{u}_1(\rho, \vartheta; f), \quad \widehat{u}_2(\rho, -\vartheta; -f) = -\widehat{u}_2(\rho, \vartheta; f).$$

Now, by linearity, we must have that

$$\widehat{u}_1(\rho, -\vartheta; -f) = -\widehat{u}_1(\rho, -\vartheta; f), \quad \widehat{u}_2(\rho, -\vartheta; -f) = \widehat{u}_2(\rho, \vartheta; f).$$

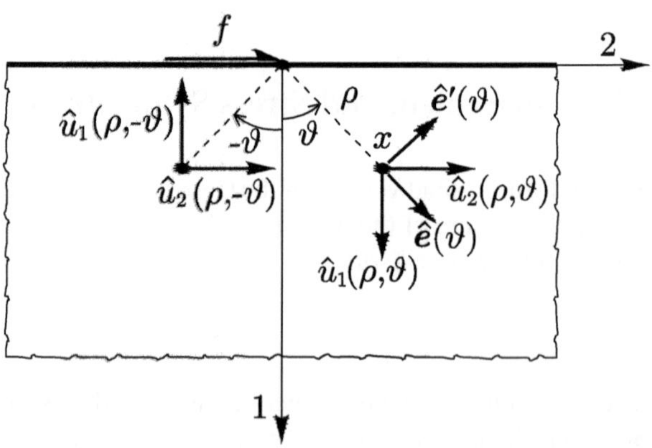

Fig. 8.2 Displacement symmetries in the Cerruti problem

In conclusion, whatever the load direction and intensity, (8.1) must hold. Alternatively, with the use of the physical basis $(\boldsymbol{e}, \boldsymbol{e}', \boldsymbol{e}_3)$, those conditions can be written as follows[1]:

$$\widehat{u}_\rho(\rho, \vartheta) = -\widehat{u}_\rho(\rho, -\vartheta), \quad \widehat{u}_\vartheta(\rho, \vartheta) = \widehat{u}_\vartheta(\rho, -\vartheta). \tag{8.2}$$

Finally, intuition suggests to expect that the Cerruti stress field $\boldsymbol{S}^C$ satisfies the following condition:

$$\widehat{\boldsymbol{S}}^C(\rho, \vartheta)\widehat{\boldsymbol{e}}(\vartheta) \cdot \widehat{\boldsymbol{e}}(\vartheta) = -\widehat{\boldsymbol{S}}^C(\rho, -\vartheta)\widehat{\boldsymbol{e}}(-\vartheta) \cdot \widehat{\boldsymbol{e}}(-\vartheta). \tag{8.3}$$

8.2 The Stress Field of the 2-D Cerruti Problem

In this section, just as we did when we dealt with the Flamant Problem, we consider the 2-D version of the Cerruti Problem, that is, the problem of a concentrated tangent load applied at a boundary point of a half-plane.

The balanced stress field for this problem obtains by picking

$$\alpha_0 = -\frac{2f}{\pi} \quad \text{and} \quad \widehat{a}(\vartheta) = \sin\vartheta \tag{8.4}$$

in representation (4.10); therefore, it has the form:

$$\widehat{\boldsymbol{S}}^C(\rho, \vartheta) = -\frac{2f}{\pi}\rho^{-1}\sin\vartheta\,\widehat{\boldsymbol{e}}(\vartheta) \otimes \widehat{\boldsymbol{e}}(\vartheta), \quad \text{for } \rho \in [0, +\infty),\ \vartheta \in [-\pi/2, +\pi/2] \tag{8.5}$$

(note that this field satisfies condition (8.3)).

Here is how we reason to justify the choices in (8.4). We saw in Sect. 4.3.2 that, for stress fields of type (4.10), the compatibility condition (2.69) translates into the request that function $\widehat{a}$ verifies a well-known ODE:

$$a'' + a = 0, \tag{8.6}$$

whose general solution in the interval $(-\pi/2, +\pi/2)$ consists in a linear combination with arbitrary coefficients of the even solution $\widehat{a}^e(\vartheta) = \cos\vartheta$ and the odd solution $\widehat{a}^o(\vartheta) = \sin\vartheta$. Moreover, in Sect. 4.3.3, we have formulated the equilibrium condition for the equilibrium of the half-disk $\mathcal{S}_\rho$, a vector condition that in the present circumstances reads:

[1] Recall that

$$u_\rho = \boldsymbol{e} \cdot (u_\alpha \mathbf{e}_\alpha) = u_1 \cos\vartheta + u_2 \sin\vartheta,$$
$$u_\vartheta = \boldsymbol{e}' \cdot (u_\alpha \mathbf{e}_\alpha) = -u_1 \sin\vartheta + u_2 cos\vartheta,$$

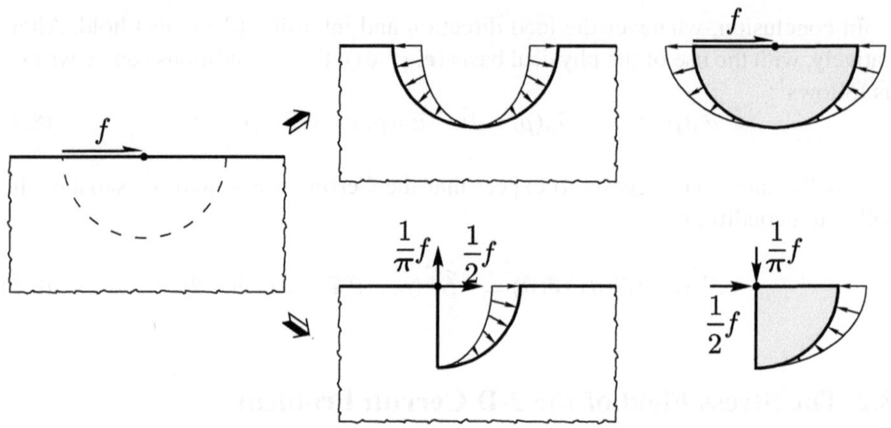

Fig. 8.3 Equilibrium of portions of a Cerruti half-plane, in the shape of a half-disk and a quarter-disk (this figure is taken from [2])

$$\alpha_0 \int_{-\pi/2}^{+\pi/2} \Big(\rho^{-1}\widehat{a}(\vartheta)(\cos\vartheta\,\boldsymbol{e}_1 + \sin\vartheta\,\boldsymbol{e}_2)\Big)\rho\,d\vartheta + f\boldsymbol{e}_2 = \boldsymbol{0},$$

and it is therefore equivalent to the scalar conditions:

$$\int_{-\pi/2}^{+\pi/2} \widehat{a}(\vartheta)\cos\vartheta\,d\vartheta = 0 \quad \text{and} \quad \alpha_0 \int_{-\pi/2}^{+\pi/2} \widehat{a}(\vartheta)\sin\vartheta\,d\vartheta + f = 0,$$

the first of which is satisfied automatically if we pick an odd solution of (8.6), as we are obliged to do in order to satisfy (8.3), while the second implies that

$$\alpha_0 = -\frac{2f}{\pi}.$$

Thus, relation (8.5) is established.[2]

Remark 8.1 In analogy to what we observed for the Flamant Problem in Sect. 4.4.2, we can use formula (8.5) to represent the contact interaction on the boundary of parts having the shape of a half-disk, when thought as isolated from their complement with respect the half-plane itself, and in equilibrium (Fig. 8.3). Once again, we see that concentrated contact interactions may be necessary for part-wise equilibrium.

Remark 8.2 A linear combination of Flamant's and Cerrruti's 2-D stress fields yields:

[2] A plane Cerruti stress field can be constructed also by an *ad hoc* use of the Airy method (see Sect. A.3.2).

$$\widehat{\mathbf{S}}_{FC}(\rho,\vartheta) = -\frac{2}{\pi}\rho^{-1}(\boldsymbol{f}\cdot\widehat{\boldsymbol{e}}(\vartheta))\,\widehat{\boldsymbol{e}}(\vartheta)\otimes\widehat{\boldsymbol{e}}(\vartheta), \quad \text{for } (\rho,\vartheta)\in[0,+\infty)\times[-\pi/2,+\pi/2].$$

This everywhere divergenceless stress field solves the equilibrium problem of a half-plane loaded by a concentrated force $\boldsymbol{f}$, whatever its relative inclination. Moreover, the stress field in a wedge $\mathcal{W}_{\vartheta_0}$ with vertical angle $2\vartheta_0$ is obtained by computing the constant α_0 in formula (4.10) so as to guarantee the equilibrium of a wedge slice of arbitrary radius; it is found that

$$\widehat{\mathbf{S}}_{W}^{FC}(\rho,\vartheta) = -\frac{1}{\vartheta_0}\rho^{-1}(\boldsymbol{f}\cdot\widehat{\boldsymbol{e}}(\vartheta))\,\widehat{\boldsymbol{e}}(\vartheta)\otimes\widehat{\boldsymbol{e}}(\vartheta), \quad \text{for } (\rho,\vartheta)\in[0,+\infty)\times[-\vartheta_0,+\vartheta_0].$$

Remark 8.3 A glance to (8.5) is enough to see that a constant-magnitude locus of that stress field is any set of points whose polar coordinates satisfy

$$\rho^{-1}\sin\vartheta = \text{a given constant.}$$

Now, a circumference of radius $|c|$ centered at point $(x_1 = 0, x_2 = c)$ has the Cartesian equation:

$$c^2 = x_1^2 + (x_2 - c)^2 = (\rho\cos\vartheta)^2 + (\rho\sin\vartheta - c)^2 = \rho^2(1 - (2c)\rho^{-1}\sin\vartheta + c^2/\rho^2),$$

or rather, in polar coordinates,

$$\rho^{-1}\sin\vartheta = (2c)^{-1}.$$

For each value of $|c|$, there are two such circumferences, shown in Fig. 8.4; the plane Cerruti stress tensor has constant magnitude at each point of these circumferences with $x_1 \geq 0$.

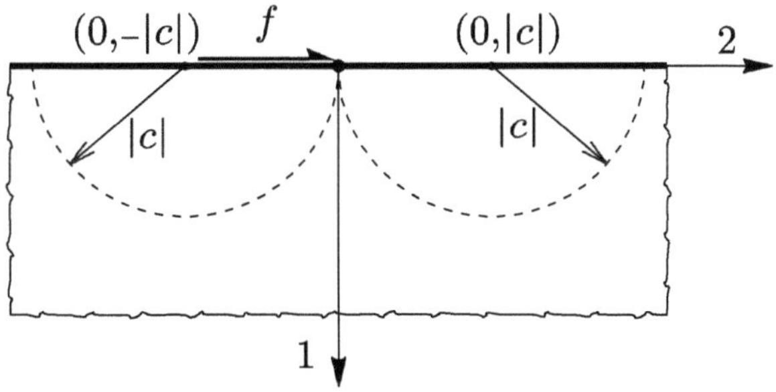

Fig. 8.4 Locus of constant stress-magnitude

8.3 The Diffused-load Cerruti Problem

8.3.1 The Stress and Strain Fields

The compatible 3-D stress field associated with the stress field (8.5) is:

$$S = \widetilde{S}^C(\rho, \vartheta, x_3) = -\frac{2f}{\pi}\rho^{-1}\sin\vartheta\Big(\widehat{e}(\vartheta)\otimes\widehat{e}(\vartheta) + \nu\, e_3\otimes e_3\Big)$$

(cf. the equally interrelated fields (4.18) and (4.14)). Again, it is quickly checked that:

- $\operatorname{div} S = 0$ at interior points of $\mathcal{HS}^+$;
- $Sn = -Se_1 \equiv 0$ on the $x_1 = 0$ plane, except over the line where the load is applied;
- $\lim_{\rho\to+\infty} \widetilde{S}^C(\rho, \vartheta, x_3) = 0$, per ogni scelta di (ϑ, x_3).

Remark 8.4 Both in Flamant's and Cerruti's problem, the dimensions of stress depend on the dimensions of the space; in their 2-D versions, the load factor f has the dimension of a force ($\dim(f) = F$), and then $\dim(S) = FL^{-1}$; in the 3-D case, $\dim(f) = FL^{-1}$, and then $\dim(S) = FL^{-2}$.

To obtain the strain field, we combine the plane stress field (8.5) with the inverse constitutive relation (2.57):

$$E = \widetilde{E}^C(\rho, \vartheta, x_3) = -\frac{2f}{\pi E_0}\rho^{-1}\sin\vartheta\Big(e\otimes e - \nu_0\, e'\otimes e'\Big).$$

As expected, the behaviour of the deformation and stress fields at infinity is the same; when ρ grows big, the displacement field tends to become rigid. Just as in the Flamant case, the change in angle of radial and circumferential material fibers is null:

$$E_{\rho\vartheta} = 0.$$

Given that

$$E_{\rho\rho} = -\frac{2f}{\pi E_0}\rho^{-1}\sin\vartheta,$$

radial fibers shorten for $\vartheta \in (0, \pi/2]$, lengthen for $\vartheta \in [-\pi/2, 0)$; moreover, since

$$E_{\vartheta\vartheta} = \nu_0\frac{2f}{\pi E_0}\rho^{-1}\sin\vartheta,$$

circumferential fibers lengthen for $\nu_0 \sin\vartheta > 0$, shorten for $\nu_0 \sin\vartheta < 0$. Finally, the change in area in the plane perpendicular to e_3 is:

$$E_{\rho\rho} + E_{\vartheta\vartheta} = -\frac{2f}{\pi E_0}\rho^{-1}\sin\vartheta(1-\nu_0);$$

therefore, the typical region represented in Fig. 4.11, becomes smaller if $0 < \vartheta \leq \pi/2$, larger if $-\pi/2 \leq \vartheta < 0$.

8.3.2 The Displacement Field

To construct the displacement field in the fashion of Sect. 4.5.3 , we have to integrate the differential system:

$$\begin{aligned} u_{\rho},_{\rho} &= -\frac{2f}{\pi E_0}\rho^{-1}\sin\vartheta, \\ u_{\vartheta},_{\vartheta} + u_{\rho} &= \nu_0\frac{2f}{\pi E_0}\sin\vartheta, \\ u_{\vartheta},_{\rho} + \rho^{-1}(u_{\rho},_{\vartheta} - u_{\vartheta}) &= 0, \end{aligned} \tag{8.7}$$

for functions $\widehat{u}_\rho$ and $\widehat{u}_\vartheta$ endowed with the symmetries (8.2).

The first equations yields:

$$\widehat{u}_\rho(\rho,\vartheta) = -\frac{2f}{\pi E_0}\ln\rho\,\sin\vartheta + \widehat{v}(\vartheta); \tag{8.8}$$

to satisfy $(8.2)_1$, $\widehat{v}$ must be odd:

$$\widehat{v}(\vartheta) = -\widehat{v}(-\vartheta).$$

From the second equation we obtain that

$$\widehat{u}_\vartheta(\rho,\vartheta) = -\frac{2f}{\pi E_0}\ln\rho\,\cos\vartheta - \widehat{V}(\vartheta) - \frac{2f}{\pi E_0}\nu_0\cos\vartheta + \widehat{w}(\rho), \tag{8.9}$$

where $\widehat{V}$ is a primitive of $\widehat{v}$, necessarily an even function. The condition determining functions $\widehat{v}$ and $\widehat{w}$ comes from the third of (8.7): by substituting into it the preliminary representations (8.8) and (8.9), we find, after some manipulations, that

$$-\frac{2f}{\pi E_0}(1-\nu_o)\cos\vartheta + v' + V = -\rho\, w' + w. \tag{8.10}$$

Now, the left side of this equation is a function of ϑ, the right side of ρ; hence, the equation itself is equivalent to two ODEs, namely,

$$(\rho\, w' - w)' = 0,$$
$$v'' + v = -\frac{2f}{\pi E_0}(1-\nu_0)\sin\vartheta.$$

The general solution of the first equation is:

$$\widehat{w}(\rho) = w_0 + w_1\rho. \tag{8.11}$$

As to the second, it is easy to verify that

$$\widehat{v}_1(\vartheta) = \frac{f}{\pi E_0}(1-\nu_0)\vartheta\cos\vartheta$$

is a solution with the desired parity; the general solution has the form:

$$\widehat{v}(\vartheta) = v_0\sin\vartheta + \frac{f}{\pi E_0}(1-\nu_0)\vartheta\cos\vartheta. \tag{8.12}$$

On taking into account (8.11) and (8.12), we find from (8.10) that

$$\widehat{V}(\vartheta) = -v_0\cos\vartheta + \frac{f}{\pi E_0}(1-\nu_0)(\cos\vartheta + \vartheta\sin\vartheta) + w_0. \tag{8.13}$$

In conclusion, with the use of (8.11), (8.12), and (8.13), the representations (8.8) and (8.9) of the displacement components are found to be:

$$\widehat{u}_\rho(\rho,\vartheta) = -\frac{2f}{\pi E_0}\ln\rho\,\sin\vartheta + \frac{f}{\pi E_0}(1-\nu_0)\vartheta\cos\vartheta + v_0\sin\vartheta,$$
$$\widehat{u}_\vartheta(\rho,\vartheta) = -\frac{2f}{\pi E_0}(1+\ln\rho)\cos\vartheta + \frac{f}{\pi E_0}(1-\nu_0)(\cos\vartheta - \vartheta\sin\vartheta)$$
$$+ v_0\cos\vartheta + w_1\rho.$$

The rigid part of this field is:

$$\widehat{\boldsymbol{u}}^C_{rig}(\rho,\vartheta; v_0, w_1) = v_0\sin\vartheta\,\widehat{\boldsymbol{e}}(\vartheta) + (v_0\cos\vartheta + w_1\rho)\widehat{\boldsymbol{e}}'(\vartheta),$$
$$= v_0\,\boldsymbol{e}_2 + w_1\boldsymbol{e}_3\times\widehat{\boldsymbol{r}}(\rho,\vartheta),\quad \widehat{\boldsymbol{r}}(\rho,\vartheta) = \rho\,\widehat{\boldsymbol{e}}(\vartheta);$$

it consists of a *translation* parallel to $\boldsymbol{e}_2$, parameterized by v_0, and a *rotation* about the coordinate axis parallel $\boldsymbol{e}_3$, parameterized by w_1. With no surprise we see that this rigid field is the general solution of the homogeneous problem associated with problem (8.7), when it is completed by the parity conditions (8.2). If we dispose of the rigid part, we remain with:

$$\widehat{u}^C_\rho(\rho, \vartheta) = -\frac{2f}{\pi E_0} \ln \rho \sin \vartheta + \frac{f}{\pi E_0}(1 - \nu_0)\vartheta \cos \vartheta,$$
$$\widehat{u}^C_\vartheta(\rho, \vartheta) = -\frac{2f}{\pi E_0}(1 + \ln \rho) \cos \vartheta + \frac{f}{\pi E_0}(1 - \nu_0)(\cos \vartheta - \vartheta \sin \vartheta), \tag{8.14}$$

the strain-inducing displacement field that solves the diffused-load Cerruti Problem.

Remark 8.5 In the same spirit as Remark 4.14, we note that the deformation mapping

$$x \mapsto y = x + \widehat{\boldsymbol{u}}^C(x) = o + (\rho + \widehat{u}^C_\rho(\rho, \vartheta))\widehat{\boldsymbol{e}}(\vartheta) + \widehat{u}^C_\vartheta(\rho, \vartheta)\widehat{\boldsymbol{e}}'(\vartheta)$$

does not enjoy local invertibility at all points of $\mathcal{HS}^+$; we return on this issue in Subsect. A.4.2 .

Remark 8.6 Instead of discarding rigid displacements altogether, we might request that the displacement of point $x = o + \rho_0 \boldsymbol{e}_1$ be null, by setting

$$v_0 = \frac{f}{\pi E_0}(1 + \nu_0 + 2 \ln \rho_0),$$

and that the rotation be null as well. We then obtain the Cartesian representation

$$\widehat{u}_1(\rho, \vartheta) = \frac{f}{\pi E_0}((1 - \nu_0)(\vartheta + (1 + \nu_0) \sin \vartheta \cos \vartheta)),$$
$$\widehat{u}_2(\rho, \vartheta) = \frac{2f}{\pi E_0} \ln \frac{\rho_0}{\rho} + \frac{f}{\pi E_0}(1 + \nu_0) \sin^2 \vartheta.$$

We can see that the displacement on the boundary of the Cerruti half-plane looks like:

$$\widehat{u}_1(x_2) = \frac{f}{2E_0}(1 - \nu_0) \operatorname{sgn} x_2,$$
$$\widehat{u}_2(x_2) = \frac{2f}{\pi E_0} \ln \frac{\rho_0}{|x_2|} + \frac{f}{\pi E_0}(1 + \nu_0).$$

References

1. Cerruti V (1882) Ricerche intorno all'equilibrio de' corpi elastici isotropi. Rend Accad Lincei 3(13):81–122
2. Podio-Guidugli P (2004) Examples of concentrated contact interactions in simple bodies. J Elast 75:167–186

Postface

In retrospect, we believe we have given our reader sufficient evidence of the advantages of the unusual integration method we systematically adopted to solve the classic problems in linear isotropic elasticity we considered, all involving concentrated loads and all formulated over unbounded domains.

Our method, we repeat, hinges on an educated use of physical intuition to guess a preliminary parametric representation for the admissible elastic states, a representation that incorporates all the symmetries dictated by a careful inspection of the body of data. Another characteristic feature of our method is that we do not formulate the problems we solve in terms of displacement. Instead, we reverse the standard order in which a solution state is sought: firstly, we try and determine, for as much as is possible, all compatible stress fields that balance the applied load; then, the corresponding strain fields; and, finally, the solution displacement field, which, as customary in linear elasticity, turns out to be unique to within, at times, an insignificant translation.

We have followed this path in the case of the two-dimensional Boussinesq-Flamant Problem, where not only the solution stress is especially easy to find—a welcomed didactic facilitation—but also turns out to be independent of material response—a rare situation. These are not the only reasons why we considered that problem first: a third reason—this one of fundamental nature—is that a careful inspection of the traction field induced by the solution stress reveals the necessity of unexpected concentrated interactions between certain adjacent body parts.

To find compatible and balanced stress fields is much harder in the case of the three-dimensional Boussinesq Problem, we now know that. But we also know that, having solved that mostly complicated problem, the two other major problems enjoying cylindrical symmetry we study, Kelvin's and Mindlin's, become relatively easy to tackle with our method, as browsing Mindlin's papers demonstrates by contrast.

P. Podio-Guidugli and A. Favata, *Elasticity for Geotechnicians*, Solid Mechanics and Its Applications 204, DOI: 10.1007/978-3-319-01258-2,

Appendix A

In this Appendix, we collect information complementary to several parts of the main text; sections are numbered according to the first occurrence of a reference to their contents.

In the first section, we recall, briefly and informally, the definitions of the absolute-value, sign, Heaviside, and Dirac, functions. In Sect. A.2, as a complement to Sect. 2.3.1, we list the constitutive equations of an isotropic linearly elastic material, in terms of both Lamé's and technical moduli.

The Airy stress function is the subject of Sect. A.3, both in general and in the forms it takes for the Boussinesq-Flamant Problem and the plane version of Cerruti Problem studied in Chap. 8.

In Sect. A.4, we take up an issue that is legitimately posed within an exact nonlinear local analysis of deformation: whether or not, given a smooth displacement field, local invertibility is guaranteed; we discuss this issue with reference to the equilibrium displacement fields solving Flamant's and Cerruti's problems.

In Sect. A.5, we record two classical representation results for Navier equation (2.40), namely, the Boussinesq-Papkovitch-Navier and Boussinesq-Somigliana-Galerkin representations. These representations are included for the sake of making our booklet reasonably complete, given that they play a central role in the classical solution of elasticity problems of interest in geomechanics. In the same spirit, in Sect. A.6, we give an exposition of the solution of Kelvin Problem, as it is presented in the old-fashioned and yet timeless classic by A.E.H. Love [5].

Section A.7, our last, contains a cumbersome computation subsuming some knowledge of complex analysis, of importance to complete the construction of the Melan stress field.

P. Podio-Guidugli and A. Favata, *Elasticity for Geotechnicians*, Solid Mechanics and Its Applications 204, DOI: 10.1007/978-3-319-01258-2,

A.1 Absolute-Value, Sign, Heaviside, and Dirac, Functions

These well-known special functions, whose common domain is the real line, are often linked the one to the other by differentiation operations that, as a rule, are to be intended in a distributional sense.

To begin with, the *absolute-value function* is defined to be:

$$x \mapsto |x|\,;$$

its derivative is the *sign function*:

$$\operatorname{sgn} x := \begin{cases} +1 & \text{for } x > 0, \\ 0 & \text{for } x = 0, \\ -1 & \text{for } x < 0 \end{cases}$$

(although defining sgn at $x = 0$ is often inessential, here a convenient definition is given). The *Heaviside function* is defined by means of the sign function:

$$H(x) := \frac{1}{2}\,(1 + \operatorname{sgn} x).$$

Finally, the *Dirac function* is the linear operator that restitutes the value in a chosen point—the origin of coordinates, say—of each given test field over the real line:

$$\delta[v] := v(0);$$

this operator may be represented by an use of the Heaviside function that identifies δ as the distributional derivative of H:

$$\delta[v] = -\int_{\mathcal{R}} v'(t)H(t)\,dt.$$

A.2 Constitutive Relations for an Isotropic Linearly Elastic Material

- *(Stress-strain laws and strain energy in terms of Lamé moduli)*

$$\begin{aligned} \boldsymbol{S} &= 2\mu\boldsymbol{E} + \lambda(\operatorname{tr}\boldsymbol{E})\boldsymbol{I}\,, \\ \boldsymbol{E} &= \frac{1}{2\mu}\Big(\boldsymbol{S} - \frac{\lambda}{3\lambda + 2\mu}(\operatorname{tr}\boldsymbol{S})\boldsymbol{I}\Big)\,, \\ \sigma(\boldsymbol{E}) &= \mu\,|\boldsymbol{E}|^2 + \frac{1}{2}\lambda(\operatorname{tr}\boldsymbol{E})^2. \end{aligned}$$

- *(Stress-strain laws and strain energy in terms of technical moduli)*

$$
\begin{aligned}
\boldsymbol{S} &= \frac{E}{1+\nu}\Big(\boldsymbol{E} + \frac{\nu}{1-2\nu}(\operatorname{tr}\boldsymbol{E})\boldsymbol{I}\Big) = 2G\Big(\boldsymbol{E} + \frac{\nu}{1-2\nu}(\operatorname{tr}\boldsymbol{E})\boldsymbol{I}\Big),\\
\boldsymbol{E} &= \frac{1}{E}\Big((1+\nu)\boldsymbol{S} - \nu(\operatorname{tr}\boldsymbol{S})\boldsymbol{I}\Big) = \frac{1}{2G}\Big(\boldsymbol{S} - \frac{\nu}{1+\nu}(\operatorname{tr}\boldsymbol{S})\boldsymbol{I}\Big),\\
\widetilde{\sigma}(\boldsymbol{E}) &= \frac{E}{2(1+\nu)}\Big(|\boldsymbol{E}|^2 + \frac{\nu}{1-2\nu}(\operatorname{tr}\boldsymbol{E})^2\Big) = G\Big(|\boldsymbol{E}|^2 + \frac{\nu}{1-2\nu}(\operatorname{tr}\boldsymbol{E})^2\Big).
\end{aligned}
$$

- *(Technical moduli in terms of Lamé moduli)*

$$
\begin{aligned}
&\textit{(Young m.)}\quad E = \frac{\mu(3\lambda+2\mu)}{\lambda+\mu},\\
&\textit{(Poisson m.)}\quad \nu = \frac{\lambda}{2(\lambda+\mu)},\qquad \textit{(shear m.)}\quad G = \mu\,.
\end{aligned}
$$

A.3 The Airy Stress Function

In 1862, George Biddel Airy (1801–1892) devised a method to construct a representation of a balanced and compatible plane stress field in terms of one scalar potential function [1].

Airy's method is fairly flexible: as we are going to show, it works for the Boussinesq-Flamant and the plane Cerruti problems, provided their different symmetries are suitably accounted for; its general traits are summarized here below.

In the absence of distance forces, the Cartesian version of the 2-D equilibrium equations reads:

$$
\begin{aligned}
S_{11,1} + S_{12,2} &= 0,\\
S_{12,1} + S_{22,2} &= 0,
\end{aligned}
\tag{A.1}
$$

two PDEs holding in a simply connected domain R. The *Airy function* $\varphi = \widehat{\varphi}(x_1, x_2)$ is a single-valued scalar function of class $C^3(R)$, such that

$$
\begin{aligned}
S_{11} &= \varphi_{,22}\,,\\
S_{22} &= \varphi_{,11}\,,\\
S_{12} &= -\varphi_{,12}\,.
\end{aligned}
\tag{A.2}
$$

It is easy to check that a stress field having the representation (A.2) satisfies the balance equations (A.1) identically; and that it satisfies identically the compatibility equation (2.69) as well, provided the Airy function is chosen *biharmonic*:

$$
\Delta\Delta\varphi = 0.
$$

A.3.1 Boussinesq-Flamant Stress

With a view toward exploiting the built-in symmetries of the Boussinesq-Flamant problem, we express the field φ in plane polar coordinates:

$$\varphi = \widetilde{\varphi}(\rho, \vartheta), \quad \widetilde{\varphi}\big(\rho(x_1, x_2), \vartheta(x_1, x_2)\big) := \widehat{\varphi}(x_1, x_2).$$

Then, by a straightforward use of $(3.19)_2$, the equation to solve for φ takes the form:

$$\begin{aligned}(\Delta\Delta\varphi =)\ &\varphi,_{\rho\rho\rho\rho} + 2\rho^{-2}\varphi,_{\rho\rho\vartheta\vartheta} + \rho^{-4}\varphi,_{\vartheta\vartheta\vartheta\vartheta} + 2\rho^{-1}\varphi,_{\rho\rho\rho} + \\ &- 2\rho^{-3}\varphi,_{\rho\vartheta\vartheta} - \rho^{-2}\varphi,_{\rho\rho} + 4\rho^{-4}\varphi,_{\vartheta\vartheta} + \rho^{-3}\varphi,_{\rho} = 0. \end{aligned} \tag{A.3}$$

Moreover, we express the components of the stress tensor in the physical basis $(\boldsymbol{e}, \boldsymbol{e}')$ in terms of the Airy representation (A.2):

$$\begin{aligned} S_{\rho\rho} &:= \boldsymbol{S}\cdot\boldsymbol{e}\otimes\boldsymbol{e} = S_{11}\cos^2\vartheta + S_{22}\sin^2\vartheta + 2S_{12}\sin\vartheta\cos\vartheta \\ &= \varphi,_2\cos^2\vartheta + \varphi,_{11}\sin^2\vartheta - 2\varphi,_{12}\sin\vartheta\cos\vartheta, \\ S_{\vartheta\vartheta} &:= \boldsymbol{S}\cdot\boldsymbol{e}'\otimes\boldsymbol{e}' = S_{11}\sin^2\vartheta + S_{22}\cos^2\vartheta - 2S_{12}\sin\vartheta\cos\vartheta \\ &= \varphi,_{11}\cos^2\vartheta + \varphi,_{22}\sin^2\vartheta + 2\varphi,_{12}\sin\vartheta\cos\vartheta, \\ S_{\rho\vartheta} &:= \boldsymbol{S}\cdot\boldsymbol{e}'\otimes\boldsymbol{e}' = -S_{11}\sin\vartheta\cos\vartheta + S_{22}\sin\vartheta\cos\vartheta + S_{12}\cos 2\vartheta \\ &= -\varphi,_{11}\sin\vartheta\cos\vartheta + \varphi,_{22}\sin 2\vartheta\cos\vartheta - 2\varphi,_{12}\cos 2\vartheta, \end{aligned} \tag{A.4}$$

where

$$\varphi,_{\alpha\beta} = \frac{\partial^2}{\partial x_\alpha \partial x_\beta}\widetilde{\varphi}\big(\rho(x_1, x_2), \vartheta(x_1, x_2)\big), \quad (\alpha, \beta = 1, 2).$$

On noting that

$$\varphi,_\beta = \nabla\varphi\cdot\boldsymbol{e}_\beta, \quad \varphi,_{\alpha\beta} = \nabla(\nabla\varphi\cdot\boldsymbol{e}_\beta)\cdot\boldsymbol{e}_\alpha = \nabla^{(2)}\varphi\cdot\boldsymbol{e}_\alpha\otimes\boldsymbol{e}_\beta,$$

with the use of (3.19), we find:

$$\begin{aligned} \varphi,_{11} &= S_{22} = \varphi,_{\rho\rho}\cos^2\vartheta + (\rho^{-1}\varphi,_\rho + \rho^{-2}\varphi,_{\vartheta\vartheta})\sin^2\vartheta - (\rho^{-1}\varphi,_{\rho\vartheta} - \rho^{-2}\varphi,_\vartheta)\sin 2\vartheta. \\ \varphi,_{22} &= S_{11} = \varphi,_{\rho\rho}\sin^2\vartheta + (\rho^{-1}\varphi,_\rho + \rho^{-2}\varphi,_{\vartheta\vartheta})\cos^2\vartheta + (\rho^{-1}\varphi,_{\rho\vartheta} - \rho^{-2}\varphi,_\vartheta)\sin 2\vartheta, \\ \varphi,_{12} &= -S_{12} = (\varphi,_{\rho\rho} - \rho^{-1}\varphi,_\rho - \rho^{-2}\varphi,_\vartheta)\sin\vartheta\cos\vartheta - (\rho^{-2}\varphi,_\vartheta - \rho^{-1}\varphi,_{\rho\vartheta})\cos 2\vartheta. \end{aligned}$$

In conclusion, relations (A.4) become:

$$\begin{aligned} S_{\rho\rho} &= \rho^{-1}\varphi,_\rho + \rho^{-2}\varphi,_{\vartheta\vartheta}, \\ S_{\vartheta\vartheta} &= \varphi,_{\rho\rho}, \\ S_{\rho\vartheta} &= -(\rho^{-1}\varphi,_\vartheta),_\rho. \end{aligned} \tag{A.5}$$

We now look for solutions by variable separation, if any, of Eq. (A.3):

$$\widetilde{\varphi}(\rho, \vartheta) = \xi(\rho)\eta(\vartheta).$$

Consistent with this supposition, we write Eq. (A.3) as follows:

$$\xi''''\eta + 2\rho^{-2}\xi''\eta'' + \rho^{-4}\xi\eta'''' + 2\rho^{-1}\xi'''\eta + \\ - 2\rho^{-3}\xi'\eta'' - \rho^{-2}\xi''\eta + 4\rho^{-4}\xi\eta'' + \rho^{-3}\xi'\eta = 0;$$

and we write Eq. (A.5) as

$$\begin{aligned} S_{\rho\rho} &= \rho^{-1}\xi'\eta + \rho^{-2}\xi\eta'', \\ S_{\vartheta\vartheta} &= \xi''\eta, \\ S_{\rho\vartheta} &= \rho^{-2}\xi\eta' - \rho^{-1}\xi'\eta'. \end{aligned} \tag{A.6}$$

For the order of singularity of all addenda to be the same in $(A.6)_1$ and $(A.6)_3$, we must choose

$$\xi(\rho) = \xi_0\rho,$$

with ξ_0 an arbitrary constant. Accordingly, the stress components become:

$$\begin{aligned} S_{\rho\rho} &= \xi_0\,\rho^{-1}(\eta(\vartheta) + \eta''(\vartheta)), \\ S_{\vartheta\vartheta} &= S_{\rho\vartheta} = 0, \end{aligned}$$

and (1.3.1) reduces to:

$$\eta'''' + 2\eta'' + \eta = 0,$$

an equation whose general solution is:

$$\eta(\vartheta) = \eta_1 \cos\vartheta + \eta_2\vartheta \cos\vartheta + \eta_3 \sin\vartheta + \eta_4\vartheta \sin\vartheta, \tag{A.7}$$

where four arbitrary constants η_i appear. For the stress field to be even in the variable ϑ, we must take $\eta_2 = \eta_3 = 0$; hence,

$$\boldsymbol{S} = S_{\rho\rho}\boldsymbol{e} \otimes \boldsymbol{e}, \quad S_{\rho\rho} = 2\xi_0\eta_4\rho^{-1}\cos\vartheta. \tag{A.8}$$

Finally, to balance the forces applied to a half-disk of contour $\mathcal{C}_\subset$ whatever its radius, that is, to have that:

$$\int_{\mathcal{C}_\rho} \boldsymbol{Se} + \boldsymbol{f} = \boldsymbol{0},$$

the constant in $(A.8)_2$ must have the value

$$\eta_4 = -\frac{f}{\pi \xi_0}, \tag{A.9}$$

so that the stress field in (A.8) takes the form (4.14).

As to the Airy function, we provisionally have:

$$\widetilde{\varphi}(\rho,\vartheta) = \xi_0 \rho(\eta_1 \cos\vartheta + \eta_4 \rho\vartheta \sin\vartheta) = \xi_0\Big(\eta_1 x_1 + \eta_4 \arctan\frac{x_2}{x_1}\, x_2\Big).$$

But, given that Airy's recipe (A.2) for stress ignores whatever linear part φ may have, we take $\eta_1 = 0$ and, by taking (A.9) into account, we eventually write the Airy function for the Boussinesq-Flamant Problem as follows:

$$\varphi^{BF}(\rho,\vartheta) = -\frac{f}{\pi}\rho\vartheta \sin\vartheta.$$

A.3.2 *Cerruti Stress*

Finding the Airy stress function for the 2-D Cerruti Problem dealt with in Sect. 8.2 is immediate. Given the prevailing symmetries, (A.7) reduces to:

$$\eta(\vartheta) = \eta_2 \vartheta \cos\vartheta + \eta_3 \sin\vartheta;$$

the correspondent stress is:

$$S_{\rho\rho} = -2\xi_0 \eta_2 \rho^{-1} \sin\vartheta;$$

moreover, again by imposing the equilibrium of a half-disk, we find that constant $\xi_0\eta_2$ must have the value f/π. All in all, the Airy function of the 2-D Cerruti problem is:

$$\varphi^{C}(\rho,\vartheta) = \frac{f}{\pi}\rho\vartheta \cos\vartheta,$$

where the linear part has been disposed of by setting $\eta_3 = 0$.

A.4 Local Invertibility of the Deformation Mapping

Let

$$\boldsymbol{u} = \widehat{\boldsymbol{u}}(\rho,\vartheta)$$

be a plane displacement field. Then, the deformation mapping is

$$x \mapsto y = x + \widehat{\boldsymbol{u}}(x) = o + (\rho + \widehat{u}_\rho(\rho, \vartheta))\widehat{\boldsymbol{e}}(\vartheta) + \widehat{u}_\vartheta(\rho, \vartheta)\widehat{\boldsymbol{e}}'(\vartheta).$$

The deformation gradient is:

$$\begin{aligned}\boldsymbol{F} = \boldsymbol{I} + \nabla \boldsymbol{u} = (1 + u_{\rho},_{\rho})\boldsymbol{e} \otimes \boldsymbol{e} + u_{\vartheta},_{\rho}\boldsymbol{e}' \otimes \boldsymbol{e} + \rho^{-1}(u_{\rho},_{\vartheta} - u_\vartheta)\boldsymbol{e} \otimes \boldsymbol{e}' \\ + (1 + \rho^{-1}(u_{\vartheta},_{\vartheta} + u_\rho))\boldsymbol{e}' \otimes \boldsymbol{e}' + \boldsymbol{e}_3 \otimes \boldsymbol{e}_3;\end{aligned}$$

its determinant is:

$$\det \boldsymbol{F} = \rho^{-1}\Big((1 + u_{\rho},_{\rho})(1 + \rho^{-1}(u_{\vartheta},_{\vartheta} + u_\rho)) - u_{\vartheta},_{\rho}(u_{\rho},_{\vartheta} - u_\vartheta)\Big).$$

We want to show how the loss-of-invertibility condition:

$$\det \boldsymbol{F} = 0 \tag{A.10}$$

looks like in the case of Flamant and Cerruti problems.

A.4.1 Flamant Deformation

The displacement field has the form (4.37). We compute:

$$\begin{aligned}1 + u_{\rho},_{\rho} &= 1 - \frac{2f}{\pi E_0}\rho^{-1}\cos\vartheta; \\ 1 + \rho^{-1}(u_{\vartheta},_{\vartheta} + u_\rho) &= 1 + \rho^{-1}\frac{2f}{\pi E_0}\nu_0\cos\vartheta; \\ u_{\vartheta},_{\rho} &= \frac{2f}{\pi E_0}\rho^{-1}\sin\vartheta \\ u_{\rho},_{\vartheta} - u_\vartheta &= -\frac{2f}{\pi E_0}\sin\vartheta.\end{aligned}$$

Hence, the condition we look for is:

$$\Big(1 - \alpha\rho^{-1}\cos\vartheta\Big)\Big(1 + \alpha\nu_0\rho^{-1}\cos\vartheta\Big) = -\alpha^2\rho^{-2}\sin^2\vartheta,$$

or rather,

$$\Big(\rho - \alpha\cos\vartheta\Big)\Big(\rho + \alpha\nu_0\cos\vartheta\Big) = -\alpha^2\sin^2\vartheta, \tag{A.11}$$

where

$$\alpha := \frac{2f}{\pi E_0} = \frac{(1-\nu)f}{\pi G} > 0 \tag{A.12}$$

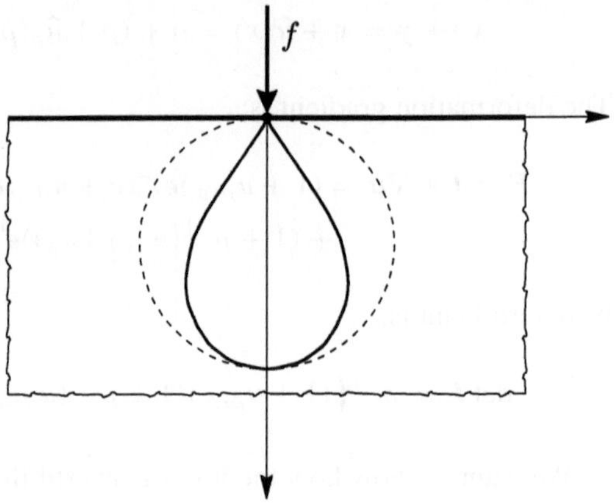

Fig. A.1 Region of local non-invertibility

and where

$$\alpha\nu_0 = \frac{\nu f}{\pi G}\,.$$

Note that the right side of (A.11) is never positive, and that

$$p_1(\rho, \vartheta) := \rho - \alpha\cos\vartheta = 0$$

is the equation of the pressure bulb of radius α. Thus, for

$$p_2(\rho, \vartheta) := \rho + \alpha\nu_0\cos\vartheta,$$

on recalling that $-1 < \nu < 1/2$, we see that,

(i) if $0 \le \nu < 1/2$, then $p_2 > 0$, and hence the *loss-of-invertibility* locus, a pear-like region, must be included within the pressure bulb ($p_1 \le 0$) (see Fig. A.1);
(ii) if $-1 < \nu < 0$, then a point (ρ, ϑ) of the loss-of-invertibility locus belongs to the pressure bulb if $p_2(\rho, \vartheta) > 0$, is external to it otherwise.

A.4.2 Cerruti Deformation

The displacement field is now given by Eq. (8.14). We compute:

$$\begin{aligned} 1+u_{\rho},_{\rho} &= 1-\frac{2f}{\pi E_0}\rho^{-1}\sin\vartheta; \\ 1+\rho^{-1}(u_{\vartheta},_{\vartheta}+u_{\rho}) &= 1+\frac{2f}{\pi E_0}\rho^{-1}\nu_0\sin\vartheta; \\ u_{\vartheta},_{\rho} &= -\frac{2f}{\pi E_0}\rho^{-1}\cos\vartheta \\ u_{\rho},_{\vartheta}-u_{\vartheta} &= \frac{2f}{\pi E_0}\cos\vartheta, \end{aligned}$$

and then the condition (A.10) reads:

$$\left(1-\frac{2f}{\pi E_0}\rho^{-1}\sin\vartheta\right)\left(1+\frac{2f}{\pi E_0}\rho^{-1}\nu_0\sin\vartheta;\right)=-\left(\frac{2f}{\pi E_0}\right)^2\rho^{-2}\cos^2\vartheta,$$

or, rather better,

$$(\rho-\alpha\sin\vartheta)(\rho+\alpha\nu_0\sin\vartheta)=-\alpha^2\cos^2\vartheta, \tag{A.13}$$

with α given by (A.12). The second member of (A.13) is never positive. and

$$p_1(\rho,\vartheta):=\rho-\alpha\sin\vartheta=0$$

is the equation of the pressure bulb of radius α. For

$$p_2(\rho,\vartheta):=\rho+\alpha\nu_0\sin\vartheta,$$

we have that:

(i) if $0\leq\nu<1/2$, then $p_2>0$, and then the loss-of-invertibility locus must be included within the pressure bulb ($p_1\leq 0$) (see Fig. A.2)
(ii) if $-1<\nu<0$, then a point (ρ,ϑ) of the loss-of-invertibility locus belongs to the pressure bulb if $p_2(\rho,\vartheta)>0$, is external to it otherwise.

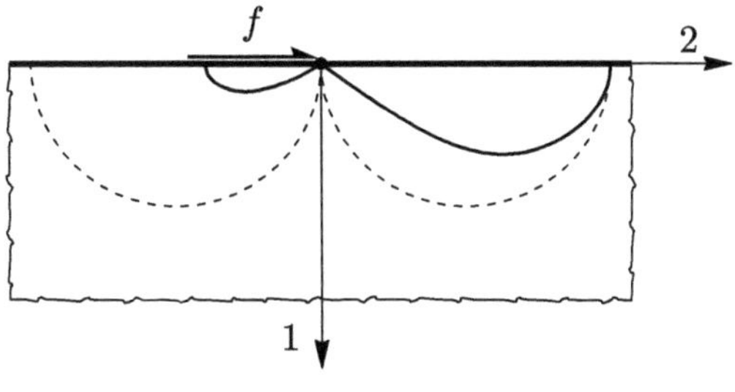

Fig. A.2 Region of local non-invertibility in Cerruti's problem

A.5 Representations of Solutions to Navier Equation

Let us write Navier equation (2.40) in the following form:

$$\Delta \boldsymbol{u} + (1-2\nu)^{-1}\nabla\big(\operatorname{div}\boldsymbol{u}\big) + \mu^{-1}\boldsymbol{d} = \boldsymbol{0}, \tag{A.14}$$

with a view towards arriving at a representation of the datum $\mapsto$ solution map $\boldsymbol{d} \mapsto \boldsymbol{u}$ in terms of suitable *potentials*, as is classical.

A possibility is to give $\boldsymbol{u}$ a *Helmholtz representation*:

$$\boldsymbol{u} = \nabla\varphi + \operatorname{curl}\boldsymbol{w}, \quad \operatorname{div}\boldsymbol{w} = 0 \tag{A.15}$$

(see [5], Chap. VIII; see also Remark 2.2). However, inserting (A.15) in (A.14) does not lead to a generally solvable system of equations for $(\varphi, \boldsymbol{w})$; the only easy-to-find representations are *incomplete*, in the sense that they hold only for certain special assignments of the datum $\boldsymbol{d}$. There are, however, at least two *complete representations* of class-$C^4(\Omega)$ solutions (see [4], Sect. 44), namely, the

- **Boussinesq-Papkovitch-Neuber representation**

$$2G\boldsymbol{u}^{BPN} = \boldsymbol{\psi} - \frac{1}{4(1-\nu)}\nabla(\boldsymbol{x}\cdot\boldsymbol{\psi} + \varphi), \quad \boldsymbol{x} := x - o, \tag{A.16}$$

where the scalar potential φ and the vector potential $\boldsymbol{\psi}$ are both of class $C^3(\Omega)$ and depend on the datum as follows:

$$\Delta\boldsymbol{\psi} = -2\boldsymbol{d}, \quad \Delta\varphi = 2\boldsymbol{x}\cdot\boldsymbol{d};$$

and the

- **Boussinesq-Somigliana-Galerkin representation**

$$2G\boldsymbol{u}^{BSG} = \Delta\boldsymbol{g} - \frac{1}{2(1-\nu)}\nabla(\operatorname{div}\boldsymbol{g}),$$

where the vector potential $\boldsymbol{g} \in C^4(\Omega)$ solves the equation

$$\Delta\Delta\boldsymbol{g} = 2\boldsymbol{d}.$$

Mindlin [6] observed that the Boussinesq-Somigliana-Galerkin representation generates Boussinesq-Papkovitch-Neuber's, provided that

$$\boldsymbol{\psi} = \nabla\boldsymbol{g}, \quad \varphi = 2\operatorname{div}\boldsymbol{g} - \boldsymbol{x}\cdot\Delta\boldsymbol{g}.$$

A.6 Kelvin Solution According to Love

(This section is taken almost *verbatim* from [3].)
In his paper [8], Kelvin does not disclose the details of the technique he adopts to solve the problem named after him: in little more than two pages, he just proposes the solution, stressing the similarity with the problem in the theory of thermal conduction, where a point *heat source* parallels the role of a *strain source* under form of a concentrated load. We now give a quick account of Kelvin problem and its solution, modeled after the account given by Love [5].[1]

Consider again Navier equation for the displacement field, this time written in the form

$$(\lambda+\mu)\nabla(\operatorname{div}\boldsymbol{u})+\mu\Delta\boldsymbol{u}+\boldsymbol{d}=\boldsymbol{0}. \tag{A.17}$$

Let the displacement and the distance force field be given a Helmholtz representation in terms of potential pairs:

$$\boldsymbol{u}=\nabla\varphi+\operatorname{curl}\boldsymbol{w},\quad \boldsymbol{d}=\nabla\psi+\operatorname{curl}\boldsymbol{b},\quad \text{with}\quad \operatorname{div}\boldsymbol{w}=\operatorname{div}\boldsymbol{b}=\boldsymbol{0} \tag{A.18}$$

(cf. Sect. A.5). Moreover, let the distance force be taken null in $\mathcal{H}\setminus\mathcal{B}_\rho$, where $\mathcal{H}$ is the whole 3-D space and $\mathcal{B}_\rho\subset\mathcal{H}$ denotes a sphere of radius ρ about the point o where the concentrated load $\boldsymbol{f}$ is applied.

On recalling that

$$\Delta\nabla(\cdot)=\nabla\Delta(\cdot)\quad\text{and}\quad \Delta\operatorname{curl}(\cdot)=\operatorname{curl}\Delta(\cdot),$$

Eq. (A.17) can be written as:

$$\nabla\big((\lambda+2\mu)\Delta\varphi+\psi\big)+\operatorname{curl}\big(\mu\,\Delta\boldsymbol{w}+\nabla\psi+\boldsymbol{b}\big)=\boldsymbol{0}. \tag{A.19}$$

A solution of (A.19) can be obtained by solving the following two equations, both set over $\mathcal{H}\setminus\mathcal{B}_\rho$:

$$(\lambda+2\mu)\Delta\varphi+\psi=0,\qquad \mu\Delta\boldsymbol{w}+\boldsymbol{b}=\boldsymbol{0}; \tag{A.20}$$

the solutions φ_ρ and $\boldsymbol{w}_\rho$ of (A.20) are:

$$\psi_\rho(x)=-\frac{1}{4\pi}\int_{\mathcal{B}_\rho}\boldsymbol{d}(\xi)\cdot\nabla_\xi(\gamma^{-1}(x,\xi))dv(\xi),$$

$$\boldsymbol{b}_\rho(x)=-\frac{1}{4\pi}\int_{\mathcal{B}_\rho}\boldsymbol{d}(\xi)\times\nabla_\xi(\gamma^{-1}(x,\xi))dv(\xi),$$

[1] Love's procedure is essentially the same adopted by those authors who solved problems of the same type of Kelvin's after him; while the flow is clear, mathematical developments are at times skipped. We warn the reader that the notation we use is rather different from the original one, and that some slight changes in presentation have been found either necessary or simply convenient.

where, for $\gamma(x,\xi) := |x-\xi|$, $(4\pi\gamma)^{-1}$ is the Green kernel of the laplacian.

The potential pair yielding the representation (A.18)$_1$ of the solution to (A.17) is found by taking the limits of φ_ρ and $\boldsymbol{w}_\rho$ for $\rho \to 0$; these limits are:

$$\varphi = \frac{1}{8\pi}\boldsymbol{f}\cdot\nabla\rho, \quad \boldsymbol{w} = \frac{1}{8\pi}\boldsymbol{f}\times\nabla\rho, \quad \rho(x) := |x-o|\,, \tag{A.21}$$

where the concentrated load $\boldsymbol{f}$ has to be intended as the limit of $\boldsymbol{d}$.[2]

Given the vector $\boldsymbol{d}$, we can determine two fields $\psi, \boldsymbol{b}$ satisfying the Helmholtz decomposition (A.18)$_2$, in two steps.

1. On applying the divergence operator to Eq. (A.18), we obtain the Poisson equation:

$$\Delta\psi = \operatorname{div}\boldsymbol{d}, \tag{A.22}$$

whose solution has the well-known representation:

$$\psi(x) = -\int_{\mathcal{B}_r} G(x,\xi)\operatorname{div}\boldsymbol{d}(\xi)\,dv(\xi), \quad x\in\mathcal{H},$$

(see, e.g., [2]), where

$$G(x,\xi) = \left(4\pi\rho(x,\xi)\right)^{-1}, \quad \rho(x,\xi) := |x-\xi|\,.$$

On recalling the identity:

$$\operatorname{div}(\varphi\boldsymbol{v}) = \varphi\operatorname{div}\boldsymbol{v} + \boldsymbol{v}\cdot\nabla\varphi,$$

and on using the divergence theorem, we obtain:

$$\psi_r(x) = -\frac{1}{4\pi}\int_{\mathcal{B}_r}\boldsymbol{d}(\xi)\cdot\nabla_\xi(\rho^{-1}(x,\xi))\,dv(\xi).$$

2. On taking the curl of $\boldsymbol{d}$, we obtain:

$$\operatorname{curl}\boldsymbol{d} = \operatorname{curl}\operatorname{curl}\boldsymbol{b} = \nabla(\operatorname{div}\boldsymbol{b}) - \Delta\boldsymbol{b} = -\Delta\boldsymbol{b}.$$

---

[2] From a classical point of view, a concentrated force is the limit of a distance force field having a shrinking support; according to the precise definition found in [7], a sequence $\{\boldsymbol{d}_n\}$ of distance force fields defined on an open neighborhood $\mathcal{R}$ of a point o tends to the load $\boldsymbol{f}$ concentrated at o if:

(i) $\boldsymbol{d}_n \in C^2(\mathcal{R})$;
(ii) $\boldsymbol{d}_n = \boldsymbol{0}$ su $\mathcal{R}\setminus\mathcal{B}_{r_n}(o)$, where $\{\mathcal{B}_{r_n}(o)\}$ is a sequence of spheres of radius r_n such that $r_n \to 0$ when $n\to\infty$;
(iii) $\lim_{n\to\infty}\int_{\mathcal{R}}\boldsymbol{d}_n = \boldsymbol{f}$;
(iv) the sequence $\{\int_{\mathcal{R}}|\boldsymbol{d}_n|\}$ is bounded.

Thus, there is another Poisson equation to solve:

$$-\Delta \boldsymbol{b} = \mathrm{curl}\boldsymbol{d};$$

its solution is:

$$\boldsymbol{b}(x) = \frac{1}{4\pi}\int_{\mathcal{B}_r} G(x,\xi)\mathrm{curl}\boldsymbol{d}(\xi)\, dv(\xi).$$

When combined with the identity

$$\mathrm{curl}(\varphi\boldsymbol{v}) = \varphi\mathrm{curl}\boldsymbol{v} + \nabla\varphi \times \boldsymbol{v},$$

an application of Stokes theorem yields:

$$\boldsymbol{b}_r(x) = -\frac{1}{4\pi}\int_{\mathcal{B}_r} \boldsymbol{d}(\xi) \times \nabla_\xi(\rho^{-1}(x,\xi))dv(\xi).$$

We now compute the limits of ψ_r and $\boldsymbol{b}_r$ for $r \to 0$, under the assumption that

$$\lim_{r\to 0}\int_{\mathcal{B}_r} \boldsymbol{d}(x)\, dv(x) = \boldsymbol{f}.$$

We find:

$$\psi(x) = -\frac{1}{4\pi}\boldsymbol{f}\cdot\nabla(r^{-1}), \quad \boldsymbol{b}(x) = -\frac{1}{4\pi}\boldsymbol{f}\times\nabla(r^{-1}), \quad r(x) := |x - o|\,,$$

or rather, since $2\nabla(r^{-1}) = \Delta\nabla r$,

$$\psi(x) = -\frac{1}{8\pi}\Delta\big(\boldsymbol{f}\cdot\nabla r\big), \quad \boldsymbol{b}(x) = -\frac{1}{8\pi}\Delta\big(\boldsymbol{f}\times\nabla r\big).$$

We can now write system (A.20) as follows:

$$\Delta\left(\varphi - \frac{1}{8\pi}\boldsymbol{f}\cdot\nabla r\right) = 0,$$
$$\Delta\left(\boldsymbol{w} - \frac{1}{8\pi}\boldsymbol{f}\times\nabla r\right) = \boldsymbol{0},$$

and arrive at the particular solution (A.21).

Finally, the displacement vector is determined by inserting (A.21) in $(\text{A.18})_1$. For $\boldsymbol{e}_2, \boldsymbol{e}_3$ two unit vectors completing an orthonormal triplet with the direction $\boldsymbol{e}_1$ of the applied load, we find:

$$u_1 = -\frac{(\lambda+\mu)f}{8\pi\mu(\lambda+2\mu)}\frac{\partial^2\rho}{\partial x_1^2} + \frac{f}{4\pi\mu\rho} = \frac{f}{16\pi G(1-\nu)}\left(\frac{2(1-2\nu)}{\rho} + \frac{1}{\rho} + \frac{x_1^2}{\rho^3}\right),$$

$$u_2 = \frac{(\lambda+\mu)f}{8\pi\mu(\lambda+2\mu)}\frac{\partial^2\rho}{\partial x_1 x_2} = \frac{f x_1 x_2}{16\pi G(1-\nu)},$$

$$u_3 = \frac{(\lambda+\mu)f}{8\pi\mu(\lambda+2\mu)}\frac{\partial^2\rho}{\partial x_1 x_3} = \frac{f x_1 x_3}{16\pi G(1-\nu)},$$

where $u_i := \boldsymbol{u}\cdot\boldsymbol{e}_i$.

A.7 Computing Two Integrals by the Method of Residues

To compute the integrals I_1, I_2 encountered in Sect. 7.2.1, we introduce the complex-variable function

$$f_1(z) := \frac{1}{(a^2+z^2)^2(x_1^2+(x_2-z)^2)^2}, \quad z \in \mathbb{C},$$

and we determine its *poles* (i.e., the points where f_1 becomes singular). We find that the solutions of equation

$$(a^2+z^2)^2(x_1^2+(x_2-z)^2)^2 = 0,$$

are:

$$z_{1,2} = \pm ai, \qquad z_{3,4} = x_2 \pm i x_1,$$

each one with algebraic multiplicity 2. It follows that f_1 can be written as:

$$f_1(z) = \frac{1}{(z-ai)^2(z+ai)^2(z-x_2-ix_1)^2(z-x_2+ix_1)^2}.$$

Let C be a closed integration path consisting of the union of interval $(-R,+R)$ and a circumference γ whose radius R is big enough to encompass the poles of $f_1(z)$ having positive imaginary part (Fig. A.3). According to the residues theorem,

$$\oint_C f_1(z)dz = \int_{-R}^{+R} f_1(x)dx + \int_\gamma f_1(z)dz = 2\pi i \sum_{j=1}^{n} (\text{Res } f_1)(z_j),$$

where n is the number of poles of f_1; this result is a consequence of the fact that $f_1(z) = O(R^{-8})$, and hence

$$\lim_{R\to\infty}\int_\gamma f_1(z)dz = 0.$$

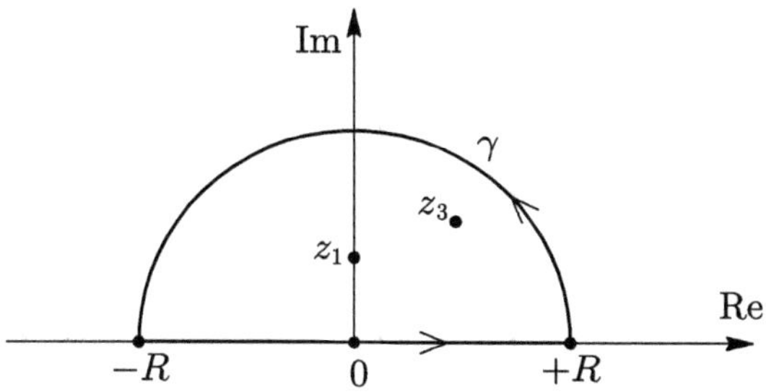

Fig. A.3 The integration path $\mathcal{C}$

Thus,

$$I_1 = 2\pi i\Big((\text{Res } f_1)(ai) + (\text{Res } f_1)(x_2 + ix_2)\Big).$$

The computation of residues yields:

$$\begin{aligned}(\text{Res } f_1)(ai) &= \lim_{z\to ai} \frac{d}{dz}\big((z-ai)^2 f_1(z)\big)\\ &= \frac{6ax_2 + i(x_1^2 + x_2^2 - 5a^2)}{4a^3(a - x_1 + ix_2)^3(a + x_1 + ix_2)^3},\\ (\text{Res } f_1)(x_2 + ix_2) &= \lim_{z\to x_2+ix_2} \frac{d}{dz}\big((z - x_2 - ix_2)^2 f_1(z)\big)\\ &= -\frac{i(a^2 - 5x_1^2 + 6ix_1x_2 + x_2^2)}{4x_1^3(a^2 - (x_1 - ix_2)^2)^3};\end{aligned}$$

and hence,

$$I_1 = \frac{2\pi}{4a^3x_1^3}\frac{(x_1 + a)^3(x_1^2 + 3ax_1 + a^2) + (a^3 + x_1^3)x_2^2}{\big((x_1 + a)^2 + x_2^2\big)^3}.$$

Quite similarly, to compute I_2 we introduce the complex-variable function

$$f_2(z) = \frac{z^2}{(z - ai)^2(z + ai)^2(z - x_2 - ix_1)^2(z - x_2 + ix_1)^2},$$

having the same poles as $f_1(z)$; then,

$$I_2 = 2\pi i\big((\text{Res } f_2)(ai) + (\text{Res } f_2)(x_2 + ix_2)\big).$$

Computing residues yields:

$$(\text{Res } f_2)(ai) = \lim_{z \to ai} \frac{d}{dz}\big((z - ai)^2 f_2(z)\big) = \frac{i(3a^2 + x_1^2 + 2aix_2 + x_2^2)}{4a(a - x_1 + ix_2)^3(a + x_1 + ix_2)^3},$$

$$\begin{aligned}(\text{Res } f_2)(x_2 + ix_2) &= \lim_{z \to x_2 + ix_2} \frac{d}{dz}\big((z - x_2 - ix_2)^2 f_1(z)\big) \\ &= -i\frac{x_1^2(x_1 + a)^3 + (x_1 + a)(x_1^2 + 5ax_1 + a^2)x_2^2 + ax_2^4}{4ax_1^3\big((x_1 + a)^2 + x_2^2\big)^3};\end{aligned}$$

consequently,

$$I_2 = \frac{\pi}{2ax_1^3}\frac{x_1^2(x_1 + a)^3 + (x_1 + a)(x_1^2 + 5ax_1 + a^2)x_2^2 + ax_2^2}{\big((x_1 + a)^2 + x_2^2\big)^3}.$$

References

1. Airy GB (1863) On the strains in the interior of beams. Phil Trans R Soc Lond 153:49–80
2. Evans LC (2010) Partial differential equations. American Mathematical Society, Providence
3. Favata A (2012) On the Kelvin problem. J Elast 109:189–204
4. Gurtin ME (1972) The linear theory of elasticity. In: Flügge S (ed) Handbuch der Physik, vol VIa/2. Springer, Berlin.
5. Love AEH (1927) A treatise on the mathematical theory of elasticity. Cambridge University Press, Cambridge
6. Mindlin RD (1936) Note on the Galerkin and Papkovitch stress functions. Bull Am Math Soc 42:373–376
7. Turteltaub MJ, Sternberg E (1968) On concentrated loads and green's functions in elastostatics. Arch Ration Mech Anal 29:193–240
8. Thompson W (Lord Kelvin) (1848) Note on the integration of the equations of equilibrium of an elastic solid. Camb Dublin Math J 3:87–89

Index

A
Airy function, 63, 152, 161, 163, 166

B
Basis
 contravariant b., 43
 contravariant b. vectors, 47, 49
 covariant b., 43
 covariant b. vectors, 47, 49
 physical b., 46, 47, 49, 164
Bessel function, 98
Boundary
 reduced b., 68
 topological b., 68
Boundary conditions
 Dirichlet-type b.c., 4, 129
 Neumann-type b.c., 4

C
Cauchy's Stress Theorem, 28
Cesàro's formula, 23
Compatibility, 22
 c. in stress terms, 38
Compatibility conditions, 22
Contact actions
 concentrated, 30, 67
 concentrated c.a., 27, 29, 67–70, 102, 103
 diffused c.a., 26
Contact interactions
 concentrated c.i., 27, 29, 152
 diffused c.i., 27
Coordinates
 curvilinear c., 43, 72, 82
 cylindrical c., 46, 57, 58, 86, 94, 96, 97, 101, 148
 orthogonal c., 44
 polar c., 44, 59, 72, 153, 155, 164
 spherical c., 48, 94, 98
Curl
 of a tensor field, 22
 of a vector field, 19

D
Dilatation modulus, 33
Dirac delta, 6, 65, 144, 162
Directional derivative, 18
Displacement, 17, 18
 d. gradient, 18
 plane d., 23
 rigid d., 20
Distance actions, 26
 inertial d. a., 26
Divergence
 d. measure, 66
 d. of a tensor field, 52
 d. of a vector field, 52
Dyadic product, 7

E
Elasticity tensor, 30
Euler cut, 29, 58, 103

F
Free-body diagram, 29, 67, 68

G
Geotechnical applications, 64, 79, 108, 133, 161
Gradient
 g. of a scalar field, 51
 g. of a vector field, 51
Green

P. Podio-Guidugli and A. Favata, *Elasticity for Geotechnicians*, Solid Mechanics and Its Applications 204, DOI: 10.1007/978-3-319-01258-2,

Green (*cont.*)
G. function, 1, 4, 6, 7, 9, 13, 79, 104, 136, 142, 143, 172
G.'s second identity, 144

H
Heaviside function, 6, 11, 162
Helmholtz's Decomposition Theorem, 21
H. representation, 170–172

I
Internal constraints, 117
incompressibility, 117
rigidity, 120
shape preservation, 120

J
Jump operator, 8, 115

K
Kronecker's symbol, 19

L
Lamé's moduli, 31, 34, 162, 163
Laplace equation, 93, 96, 97
Laplacian, 31
Local invertibility, 157, 161
loss-of-invertibility locus, 168, 169

N
Navier equation, 31, 32, 35, 113, 161, 170, 171
Neuman function, 98

O
Oedometric conditions, 79

P
Poisson
P. equation, 173
P. modulus, 33, 118
Pressure bulb, 108, 109, 168, 169

R
Reciprocity Theorem, 5
Residues, 174
Ricci's symbol, 19
Rotation tensor, 19

S
Shear modulus, 33
Sign function, 9, 12, 75, 161, 162
Signorini's Lemma, 97
Strain, 17
infinitesimal s. tensor, 20
plane s. field, 24
s. energy, 30, 31, 34, 35, 71, 162, 163
s. nuclei, 9
s. tensor, 19
Stress, 28
active s., 120
plane s., 36
reactive, 120
s. tensor, 28
s. vector, 27
universal s. field, 39
Strong formulation, 4

V
Virtual velocity, 121

W
Weak formulation, 11, 30, 65
Well-posedness, 22
Working
contact w., 66
distance w., 66
load w., 121
of a concentrated force, 67
stress w., 66, 121

Y
Young modulus, 33

MIX
Papier aus verantwortungsvollen Quellen
Paper from responsible sources
FSC® C105338

If you have any concerns about our products,
you can contact us on
ProductSafety@springernature.com

In case Publisher is established outside the EU,
the EU authorized representative is:
Springer Nature Customer Service Center GmbH
Europaplatz 3, 69115 Heidelberg, Germany

Printed by Libri Plureos GmbH
in Hamburg, Germany